The common fields of England

by the same author

Trade and banking in early modern England

Textile manufactures in early modern England

The farmers of old England

Agrarian problems in the sixteenth century and after

The agricultural revolution

To Ann

ERIC KERRIDGE

The common fields of England

Manchester University Press
Manchester and New York

distributed exclusively in the USA and Canada
by St. Martin's Press

Published by Manchester University Press
Oxford Road, Manchester M13 9PL, UK
and Room 400, 175 Fifth Avenue, New York, NY 10010, USA

Distributed exclusively in the USA and Canada
by St. Martin's Press, Inc., 175 Fifth Avenue, New York, 10010, USA

British Library Cataloguing-in-Publication Data
A catalogue record for this book is available from the British Library

Library of Congress cataloging in publication data
Kerridge, Eric.
The common fields of England / Eric Kerridge.
p. cm.
Includes bibliographical references and indexes.
ISBN 0-7190-3572-4 (hardback)
1. Commons—England. I. Title.
HD 1289.G7K47 1992
333.2—dc20 91-38792

ISBN 0 7190 3572 4 *hardback*

Printed in Great Britain
by Biddles Ltd, Guildford and King's Lynn

Contents

Abbreviations

Acc.	Accession
Acq.	Acquisition
A.O.	Augmentation Office
Arch.	Archdeaconry
B.A.	Bulk Accession
Ch.	Charter
Cons.	Consistory
Dep.	Deposition
D.L.	Duchy of Lancaster
Ecton	Sotheby (Ecton)
F-H	Finch-Hatton
F(M)	Fitzwilliam (Milton)
G'bury	Gorhambury
G.S.	General Series
Hood	Gregory-Hood
I(L)	Isham(Lamport)
Lans.	Lansdowne
Leigh	Stoneleigh (Leigh)
L.R.	Land Revenue
M.B.	Miscellaneous Book(s)
Mont.	Montagu
M.R.	Manorial Rolls
Pec.	Peculiar
R.	Roll(s)
R.&S.	Rentals and Surveys
Req.	Court of Requests Proceedings
S.P.D.	State Papers Domestic
St.Ch.	Star Chamber
T.R.	Treasury of Receipt
W. de B.	Willoughby de Broke
Wds	Court of Wards and Liveries

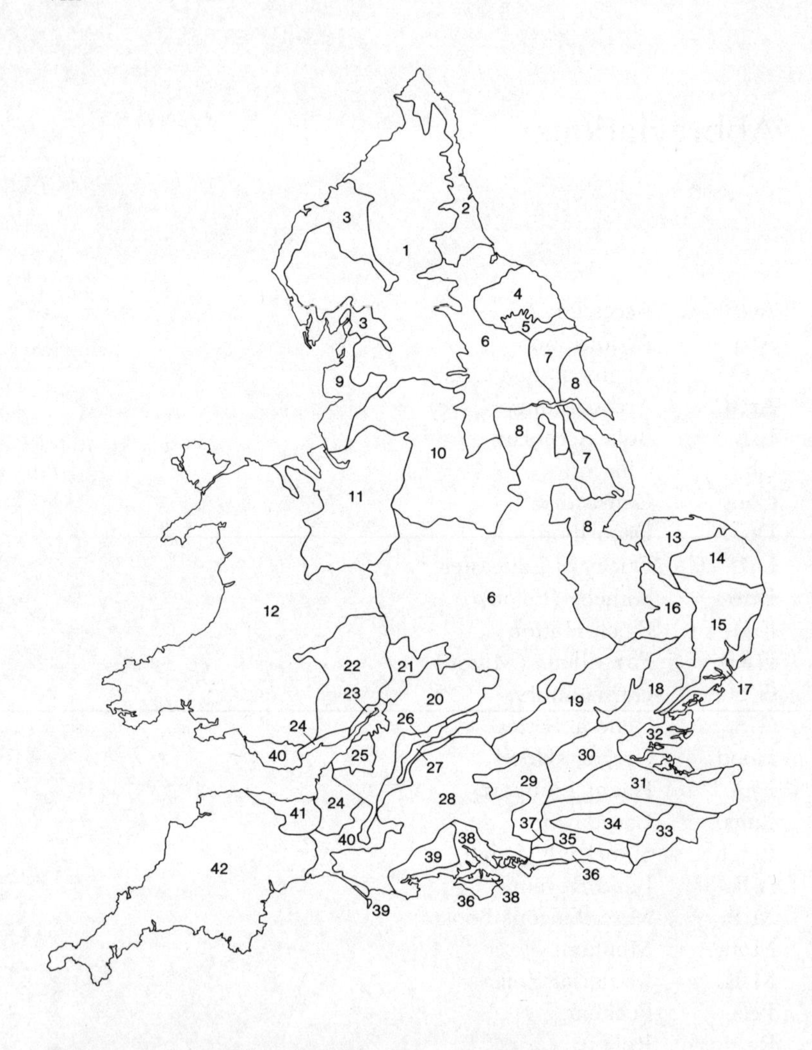

The farming countries of early modern England
1 North Country **2** North-eastern Lowlands **3** North-western Lowlands **4** Blackmoors **5** Vale of Pickering **6** Midland Plain **7** Northwold Country **8** Fen Country **9** Lancashire Plain **10** Peak-Forest Country **11** Cheshire Cheese Country **12** Wales **13** Norfolk Heathlands **14** East Norfolk **15** High Suffolk **16** Breckland **17** Sandlings Country **18** Woodland **19** Chiltern Country **20** Cotswold Country **21** Vale of Evesham **22** Vales of Hereford **23** Vale of Berkeley **24** Western Waterlands **25** Bristol District **26** Cheese Country **27** Oxford Heights **28** Chalk Country **29** Blackheath Country **30** Vale of London **31** Northdown Country **32** Saltings Country **33** Romney Marsh **34** High Weald **35** Wealden Vales **36** Southdown Country **37** Petworth District **38** South Seacoast Country **39** Poor Soils Country **40** Butter Country **41** Vale of Taunton Deane **42** West Country

Introduction

A common field is one in which various parts or parcels of land (or the use of them) belong to individual proprietors, who exercise sole proprietary rights when the land is in crop but leave them in abeyance when it is not, so that when not in crop, the land is under the general management of all the proprietors in common and by common agreement. Thus a common field is alternately closed for cropping and thrown open to all the commoners, that is, to all the owners of common rights. A common meadow is a common field in which hay is the only crop to be cut. A common pasture is one in which the various proprietors have common rights of pasture. It goes without saying that common fields, meadows, marshes, pastures, heaths, woods, and all commons whatsoever are common only to the proprietors and to no one else.

Hundreds of years ago common fields were a familiar part of the English landscape. Now they belong to the past. We take it for granted that farm fields are in single ownership and enclosed by hedges, fences or walls. Over a large part of the Continent, however, common fields are still a familiar sight and taken for granted. Ask a Tyrolean why the fields in his part of the world are divided into such small pieces and he is more than likely to answer, that it is due to division between heirs; but then he will not be meaning to explain the existence of common fields as such, for these he takes to be an unremarkable part of the normal state of things; he will merely mean to explain why the parcels in the fields are as small as they are. He cannot envisage permanent arable not being in common fields, and would wonder at the enclosed arable fields of England.

In recent times many more English people have had the opportunity of seeing common fields. They are unlikely to have seen any of the remnants of the ones in Laxton in Nottinghamshire, in the Isle of Portland and one or two other places in this kingdom, but many have seen, even if with unseeing eyes, some of the Continental common fields. Soldiers stationed in Germany may have seen common fields and common meadows in such places, for example, as Berleburg in Kreis Siegen. Anyone who has flown over Europe and has peered down at the land below will have seen unfamiliar field patterns: wide, open fields, often divided into strips of land distinguished by the variety of colours and shades arising from the different crops within them. These are common fields. To see common fields, the holidaymaker in Innsbruck has only to go up to the Hungerburg and look down, or along the Stubaital to Mutters or Natters, or to Lermoos in Loisachtal. In Salzburg, he has only to walk a few miles down the Salzach valley towards Elsbethen and stray a few yards from the main road. In Davos, he need only climb some way towards the Weissfluhjoch and look back, and if he then walks across to Sertig Dörfli and saunters down to the river, his path will lead him into a common meadow complete with the wooden stakes that, in these parts, usually divide the parcels one from another. Many other holiday haunts in Europe and beyond offer similar views.

This book is primarily concerned with an historical investigation of English common fields, their origins, practices, characteristics, and, above all, their historical significance.

Historians delving into the earliest history of agriculture inevitably encounter the work of archaeologists who have unearthed artefacts and cultivations they assign to approximate dates that straddle the indeterminate line between history and prehistory. Many archaeologists have greatly added to the sum of knowledge, but strictly archaeological discipline is incapable, by itself, of explaining the artefacts discovered by archaeologists. Agricultural implements, for example, can be described in purely physical terms, but cannot be so explained. This can only be done, if at all, by drawing on our knowledge of a wide range of subjects, including Man's ways of reasoning and acting, and, here especially, agricultural matters such as soils, subsoils, climature and systems of general management. Such knowledge is accessible equally to both archaeologists and historians. It follows that historians may be just as capable of explaining an artefact faithfully described by an archaeologist as are the archaeologists themselves. Once the physical characteristics and provenance of an artefact correctly assigned to historical times have been demonstrated as fully and accurately as possible, the historian and the archaeologist are on equal terms in attempting to explain its significance, and whether the one or the other succeeds in providing the better explanation will depend entirely on the knowledge, perspicacity and percipience of the individual.

Dating is the archaeologists' difficulty. Every means they may use to date the fields and other artefacts of interest to us here, is imperfect: pollen analysis, because pollen is so easily and frequently moved; typology, because techniques, arts and fashions were unevenly diffused both geographically and chronologically; and carbon-14, because the amount of this in the atmosphere is not always constant. More generally, the haphazardly varied fortunes of discarded artefacts remain largely unknown. Consequently, many archaeological datings of artefacts are suspect and some contested by archaeologists themselves. At best, most archaeological datings are somewhat imprecise, and so of limited help to historians.

Happily, in the course of this present study, little or no disagreement has arisen between the findings of historians and those of archaeologists. Almost the only disagreements met with have been between archaeologist and archaeologist or between historian and historian.

The earliest times apart, the period covered in this work has been dominated by historians. Historical investigations of common fields in north-western Europe have passed through at least two main phases. In the first, it tended to be assumed that common fields had come into being throughout the region during the Dark Ages and more or less simultaneously. A second phase opened with the discovery by German and other Continental scholars that common fields emerged in their countries only long afterwards, in the high Middle Ages or early modern times, or, in parts of Yugoslavia and elsewhere, not until the nineteenth century. In England these findings prompted the proposition that here too the development of common fields chiefly occurred in the high Middle Ages and the early modern period. What we attempt to show below is that this development came early in England and only much later on the Continent, suggesting that permanent common fields of types suitable to north-western Europe were first invented by the English.

At one time it was claimed that the Germanic peoples were the first to introduce ridge-and-furrow ploughing with a plough equipped with a coulter, share and mould-board, and that this gave rise to common fields. When it was later discovered that the introduction of such ploughs had not immediately resulted in the formation of common fields, the significance of this introduction tended to be discounted. We argue, however, that it paved the way for the eventual development of permanent common fields and was, moreover, in its own right, an innovation of the greatest importance, for ridge-and-furrow ploughing was essential to permanent cultivation in all the heavier and wetter lands of north-western Europe. Tiresome and ignoble as the technique of ploughing may seem to some, it was none the less crucially important in the early history of England. Without the ridge-and-furrow method of ploughing, the growth and development of our kingdom would have been severely stunted.

Past historians have between them put forward a number of theories to explain the origins of common fields. Some have found their cause to lie in the type of plough employed, some in partible inheritance, some in a rise in the population, some in co-aration with common plough-teams, some in the planned distribution of land. We prefer to argue that the division of land into common fields arose from a variety of different causes in different places and that greater stress should be put upon the *causa causarum*, the fundamental cause of common fields, namely, the division both of the land itself and, just as importantly, of the livestock, among a mass of small occupiers. Conversely, the supersession of a multitude of small occupiers by a small number of large ones, sounded the death-knell of common fields.

In broad terms, the general effect of our argument is to overturn two previous theories. First, the notion that common-field development was ubiquitous and simultaneous is replaced by the one that it started in one particular heartland and then gradually spread in wider and wider circles, as the ripples on a pond from the place into which a stone has been thrown. Secondly, we reject the old assertion that this diffusion proceeded from east to west, and argue that it went mainly from west to east, from England right across to Russia.

CHAPTER 1

The characteristics of English common fields

A common field was a tract of land subject to common rights of pasture except when in crop and then necessarily fenced against stray animals, kept several to the individual cultivators and debarred to all other commoners. The essence of a common field was in these common rights. Properties and holdings in common fields were usually dispersed and intermixed, but this intermingling or intermeddling was inessential. Nor was joint interest involved, for there were many half-year meadows, where one man had the hay and another the aftermath, that were not common meadows and not subject to common rights.[1] Common fields were often known as town fields, especially in the north-west, the public body concerned being the town. Yet even where a degree of management pertained to the town, ownership was private and individual. In short, what distinguished common fields from land in severalty was nothing but the existence of common rights.

Despite the formation within them of Lammas closes, which were still subject to common rights, common-field land was largely open and unenclosed by any permanent mound or bound, a temporary fence or dead hedge being made or completed as required. But 'common field' and 'open field' are not synonyms, for not all common fields were open nor all open ones common. Hence nothing but disadvantage attaches to the current fashion of employing the term 'open field' for what was formerly known as a 'common' or 'town' field. Two distinctions need making: between open and enclosed fields and between common lands and severalty.[2]

Common fields were largely in tillage and it was ploughing that gave them many of their characteristics. The fields were mostly divided into furlongs, as cultivation dictated. Except where the land was flat, which it seldom was, one had to plough this way or that according to the lie of the land in order to get the best natural drainage and avoid soil erosion. And even where the land was flat, horse sense and human reason often led the ploughman to moderate the length of his furrow, for the shorter the furrow, the easier to keep it straight. In drawing a furrow, the ploughman has to be able to see his poles at each end. For these and similar reasons, cultivators usually built up fields composed of numerous furlongs instead of having one enormous furlong or field. At the same time, in order to minimise the number of times the team had to be turned round, ploughmen generally tended to make their furlongs somewhat longer than wide. Had some other method of ploughing been adopted, common fields might have looked rather different, but, in fact, like those in severalty, they were mostly ploughed by inverting furrow-slices by means of a mould-board, and usually by ridge-and-furrow techniques, which are not to be confused with the formation of narrow ridges by means of digging-sticks or ridging ploughs.[3]

Ridge-and-furrow ploughing took many forms, of which the more important were gathering up from the flat, crown-and-furrow ploughing, casting, yoking or coupling ridges, slitting, throwing, splitting or cleaving down ridges, and ploughing two-in-and-two-out. Ridges were made up of inverted furrow-slices laid side by side. (These slices were commonly in one of three forms: rectangular, crested or wide. Crested ones, in particular, were employed in winter ploughing.) The middle of the ridge was called the 'crown', its two sides the 'flanks', the divisions between ridges 'open furrows' or simply 'furrows', and the edges of the furrow-slices next to the open furrows 'furrow brows'. In some open furrows, the last slices ploughed were known as 'mould' or 'hind-end' furrows. As water draining through and under the furrow-slices found its way to channels formed by open furrows, these were frequently deepened and moulded to assist surface drainage and then called 'water-furrows'. In southern parts of England, ridges were often called 'stitches', in eastern parts 'warps', and towards the north and west 'butts' or 'loons', and the intervening furrows 'reins'.[4]

Ridges were sometimes ploughed to lie north and south and receive the sun equally throughout the day; but the overriding consideration, save upon hillsides, was to make the furrows follow the slope of the ground, so as to allow water to flow more freely from the surface of the field. The width of the ridges, the frequency of the intervening furrows, and the shape of both in profile, were varied in order to provide the surface drainage demanded in particular places.[5]

The first process in ridging up land from the flat was drawing or striking the furrows. In the absence of convenient natural landmarks, a pole had to be placed in the ground at each end of the intended furrow. In drawing a furrow thus marked out, the ploughman went successively along the field sides of the two intended headlands, turning the furrow-slice outwards, in order to mark off the width of the headlands themselves. Next, the width of a quarter of a ridge was marked off to the side of the field. After this, furrows were drawn at a distance of the width of one ridge and a quarter from the first marking of the quarter ridge. When the drawing poles had been correctly aligned and drawing furrows turned, the ploughman could proceed to lay another slice against that of the drawing furrow, to form the crown of a ridge. By ploughing round and round such crowns, whole ridges were made and open furrows left between them, this being called gathering up from the flat. The furrow-slices of one half of a ridge were inclined against those of the other half, forming a crown in the middle and open furrows on the flanks. These open furrows were flat, but somewhat below the original level of the field. Were a water-furrow needed, the plough went along the open furrow, turning over a triangular slice and covering the exposed furrow brows with these new mould furrows. The resulting water-furrow was deeper and rounder than the open furrow it replaced.

Crown-and-furrow ploughing was applicable to land already gathered from the flat, as much arable was. No drawing was needed because the old furrows served the purpose. In crown-and-furrow ploughing, the open furrows became the crowns of new ridges and the crowns of old ridges the new open furrows, the whole ridge being thus shifted sideways by the space of one half ridge. The effect was to preserve the same form and elevation set up at the first gathering up from the flat, only with the ridges and open furrows transposed. This method was, therefore, a good one to employ where little water was likely to stand on the surface, where slightly elevated ridges and intervening open or water-furrows sufficed to remove any water not absorbed by soil, subsoil and base.

Gathering up from the flat raised the ridges above the former level of the field and slightly more above the bottom of the open furrows. If higher ridges were needed, each ridge was ploughed again in the same manner as before, that is to say, it was twice gathered. An open track was laid along the crown before the first two slices were turned against each other, to allow the ploughman to strike them without them overlapping and to avoid an unsightly and shoulderless crown. A ridge once gathered acquired a slight roundness at the flanks by the harrowing down of furrow brows. In a ridge twice gathered, this form became exaggerated, partly by the harrowing, but more from the want of soil at the furrow brows during the second ploughing, for this caused the outward slices to be shallower than the others. Ridges gathered twice were thus remarkably elevated and rotund. Twice-gathering was only practised in retentive soils, with the object of raising the soil above the cold, wet subsoil and of ploughing the land, so to speak, out of the water. These soils had to be ploughed so as to lie as dry as possible in autumn and winter. Wheat, in particular, needed to be kept dry and warm and the narrowest and highest ridges were ploughed for this crop. Heavy wheat lands were often gathered into high, narrow ridges, simply by repeating the same process, by thrice-gathering.

Slitting down the ridges was the exact opposite of gathering them up. It was chiefly employed in fallow stirrings and in preparing seed-beds on ridges already elevated two or three times. The furrow-slices were ploughed precisely in reverse and by half ridges, simply undoing what had been done when the ridges were gathered up. The land was laid flat again and the open furrows closed. When retentive soils were slit down over winter, therefore, gore furrows had to be ploughed between the half ridges to keep them apart and allow the water to run off into the ditches.

In coupling the ridges, all the furrow-slices of a single ridge were laid over in one direction, but in opposite directions in alternate ridges, leaving an open furrow only between each pair and throwing two ridges into one. Coupling was used to keep the land level, but was only convenient in warm, dry soils, and even here it was often necessary to make a gore furrow between the coupled ridges to serve instead of the obliterated open furrow.

Two-in-and-two-out ploughing was also employed to keep the land flat. Here the furrow-slices of two former ridges were all laid in the one direction and those of the next two ridges in the opposite direction. Four ridges were thus made into one and the open furrows reduced from six to two. This method was consequently suited only to light soils on absorbent bases in drier climates.

As an alternative to ridge-and-furrow methods, ploughing could be done in breaks or divisions without any open furrows or ridges, all the furrow-slices being laid in the one direction. But this was possible only in the warmest soils in the driest parts of the kingdom. Another possible alternative was to take the plough round and round the field. This served well enough for ploughing in green manure, but made the land rough where plough and team changed direction, leaving a mill-sail pattern in the soil. As it greatly impeded surface drainage, ploughing in mill-sail fashion was rarely practicable in severalty and unknown in common fields. Some enclosures, too, were, on occasion, ploughed rainbow fashion, following the line of a curving hedge. Finally, warm, dry, shallow soils on highly absorbent bases, like some on the higher parts of chalk downs, were merely scratched with a drag-plough or drag and not ploughed with the usual mould-board plough or sull.

The different methods of ploughing could thus be described in 1598 in the following terms:

> The manner of ploughing land is in three formes: eyther they be great lands, as with high ridges and deepe furrowes, as in all the north parts of this land, and in some sotherne parts also, or els flatte and plaine, without ridge or furrow, as in most parts of Cambridgeshiere, or els in little lands, no land containing above two or three furrowes, as in Midlesex, Essex and Hartfordshiere.[6]

In warm, dry soils on absorbent bases, where the minimum of surface drainage sufficed, no water-furrows were made; either the ridges were gathered up once and then ploughed crown-and-furrow, or they were coupled or laid two-in-and-two-out, or all the furrow-slices were laid in the same direction. In the light soils of the Chalk Country, William Cobbett found 'no ditches, no water-furrows, no drains, hardly any hedges, no dirt and mire, even in the wettest seasons'.[7] Soils on all retentive bases, however, were elevated twice or thrice into high, round ridges divided by deep water-furrows. 'If the lands lie round, the corn will not be drowned.'[8] In the north and west of the kingdom, too, where rainfall was excessive, and often the soils would not dry off quickly, the ridges were gathered two or three times.[9] High ridges were also needed in fens subject to occasional drowning. Ridges about three yards wide were gathered up as much as five or six times, so that if the worst came to the worst the crops could grow on long, narrow islands.[10] Narrow elevated ridges were especially ploughed for wheat, to keep the crop dry in winter, in all stiff, retentive soils, even in the driest parts of eastern England. Moreover, in some places, where the land was not otherwise ploughed by the ridge-and-furrow method at all and the furrow-slices were usually laid all one way in breaks or divisions, and even in the sandy loams of East Norfolk, the wheat land was nevertheless ploughed ridge-and-furrow and gathered up from the flat.[11]

Irrespective of subsoils or bases, general plans of management sometimes entailed special ploughing methods. Some excessively thin soils were set up high to deepen the seed-bed.[12] In common fields on cold land, ridges were often ploughed extremely high and sometimes very wide.[13] This was because the pulverisation brought about by many fallow stirrings deprived the soil of what little self-draining properties it had ever had, so that the first downpour of rain turned it into a quagmire. In up-and-down husbandry on the same land, however, where summer stirrings were eschewed and the old turf was preserved in the ground throughout the tillage period, soil structure was changed in the opposite way and self-draining enhanced, so the ridges did not need to be so high and sharp and were usually gathered no more than twice.[14] But it should not be supposed that highbacked ridges were ploughed only in common fields; they were sometimes continued in the succeeding enclosures.[15] Nor were 'highbacks' always ploughed in cold common fields, for when new crops, improved courses and better drainage were introduced in them, the ridges were often somewhat lowered.[16]

Land was thrown into elevated ridge and furrow not merely for the sake of tillage, but also when laying down to grass, broad, rounded ridges being then often preferred. In some districts with cold soils, lease covenants stipulated that closes be left in such ridges when laid down at the end of the term.[17] 'Highbacks' were useful for laying newly cut sheaves on, for throwing fodder on, for providing sheep and cattle with some dry pasture even in the wettest weather, for sheltering sheep from bleak winds, for increasing the surface area of grass, and for helping to weather down the soil.[18] Still another purpose of setting up high ridges with deep furrows was the floating of watermeadows on the ridge-and-furrow system.[19]

As a result of ridge-and-furrow methods of ploughing, it was often possible to discern land that had been under the plough and distinguish it from 'plains' that had not. It followed that ridge and furrow were sometimes used as circumstantial evidence for or against common rights, where both parties agreed there had been no tillage unless in common or unless in severalty.[20] Even if it were not known from documentary records that most of the closes in Dishley Grange Farm were up-and-down land in 1669 and in Bakewell's time, the evidence of their former cultivation was still on the ground in 1952.[21] Indeed, in heavy soils especially, high ridges and deep furrows were, and often still are, prominent features of the landscape. They are said to have provided cover for troops in the Battle of Naseby. And not a few droll stories are told, for instance, of the farmer who visited a neighbour busy ploughing, but could see none of the half-a-dozen teams at work because 'they were making up their furrows, and were wholly hid, by the ridges, from his sight'.[22]

Had surface drainage begun and ended with ridge-and-furrow, any surplus water would have flowed from the main body of the field or furlong only to drown its extremities. Therefore water was carried from the water-furrows and over the headlands along gutters, into main trenches or ditches, in severalty[23] and common fields alike. In common fields, guttering ('griping') and trenching were obligatory, supervised by haywards or other field-officers[24] and often provided as public services by the town government on payment of a rate.[25] Grasslands, whether in common, in severalty, open or enclosed, were often drained in much the same way with furrows and gutters.[26] Guttering could simply be done with a trenching spade or gouge or by means of a hand trenching plough;[27] but these were primarily for putting the finishing touches to the work of a great trench plough that could be operated only by individuals who maintained more than one ordinary plough team,[28] or by the whole township, as a public undertaking. Common fields were often guttered, trenched and drained by a common or public draining plough. The gutters, which extended to the very limits of the furlong, had to be at least as deep as the water-furrows and to continue in much the same line. All headlands did not need to be guttered, but where the furlong was more or less flat or convex, guttering could hardly be avoided at either end.[29]

It follows from what has been said that the ridges and furrows in former common fields cannot be distinguished visually from those in severalty. Enclosed arable fields in severalty were usually divided or grouped into shifts for field-courses, and while two or more closes often fell into one shift, two or more shifts sometimes shared the same close.[30] Moreover, even where close and shift coincided, the enclosed field was not necessarily all in one furlong. In laying down 'pastures' that had been 'corned', Blith advises the farmer, if need be, to cast each 'pasture' into several furlongs; and some enclosures were, indeed, thrown into two or more furlongs.[31] Furthermore, to prove that a field has been divided is in no way to prove that a field formerly common has been divided into enclosures, so to prove ridge and furrow have been broken by a subsequent division is not at all to prove that the ridges had been ploughed in a common field. That the subdivision of enclosures into two or three, or even as many as nine or ten closes, was a common practice from the sixteenth to the early nineteenth century, both in the Midland Plain and elsewhere, is attested by a considerable volume of evidence.[32] Such subdivision is often apparent on the ground and in field plans.[33] Throwing two or more closes into one, which was also frequent, resulted in a visual effect not unlike that produced by casting an enclosure into two or more furlongs.[34] Finally, the legibility of that palimpsest called the landscape has been still further decreased by such practices as digging up the headlands and casting their mould on the rest of the field in order to enrich it, and by the enclosure and addition to adjacent fields of roadside wastes and even the roads themselves.[35]

Let us, however, not overemphasise the extent of tillage in common lands, for within their general framework were found many and spacious common downs, heaths, wolds, hills, greens, plains, and other grasslands, not to mention common meadows. Much common-field land, too, was given over to ways and roads and to balks left unturned by the plough. Grass balks bounded severalty from common field as well as individual common fields, furlongs and parcels. The existence of grassy balks is, in general, widely known. What deserves closer attention is the internal boundary balk.

The intermingled parcels belonging to various owners and occupiers in common fields were usually divided one from another by greensward balks, locally also called 'linches', 'landsheres', 'meres', 'green furrows', 'acre reins', and so on, and often distinguished from other balks by the adjectives 'narrow' or 'foot'. These boundary balks might be between any two ridges, but were usually only between neighbour and neighbour, that is between the land of one man and that of another. Boundary balks were sometimes wider between roods than between acres, but generally ranged from two to sixteen feet in width. In heavy lands, the wider of them could not be left flat without impeding surface drainage and were, therefore, usually made in the form of narrow ridges and then often called 'green furrows'. Whatever their form, boundary balks were mown for hay when the field was in corn, and staked, tethered or baited off to horses or fed by sheep when it was in fallow or stubble. All field balks were used in this way, as well as for cart-tracks, footpaths or bridleways, but the boundary balks were as several to the different occupiers as were the parcels they bounded.[36] Dividing balks were advantageous in common-field husbandry because they provided much needed permanent grass, permitted after-cultivations with less risk of damaging the crops and poaching the land, and acted as buffers between the lands of neighbouring occupiers, so reducing trespasses. For these reasons, boundary balks were kept in permanent grass and it was generally forbidden to plough them up.[37] The parcels divided by balks were sometimes split up between two or more men and sometimes two or more parcels were consolidated into one. In the latter event it was usually permissible to plough up redundant boundary balks, but not those between neighbour and neighbour.[38] When common fields were, by general agreement, consolidated into larger parcels and redistributed, as sometimes they were, old boundary balks were often ploughed up and new ones perhaps laid down. In later times, with the introduction of new crops, some balks were ploughed up for turnips or clover.[39] Generally, the more the parcels were consolidated, the fewer the boundary balks and they disappeared altogether from some later common fields.[40] But, generally speaking, boundary balks were essential in common fields. They were left between the parcels in temporary common fields in shifting cultivation and were purposely made in permanent common fields newly laid out in the eighteenth and nineteenth centuries.[41] Balks were not the only means of marking bounds in common fields; in the wettest fenlands, goats or ditches took their place. Merestones were also widely used, especially at joints and corners and to delimit furlongs, fields, shifts and mowing doles,[42] but in the main they only supplemented the balks. Enclosures, too, had their balks. When common lands were divided up and allotted in severalty, balks were often left between the allotments, temporarily until walls were built or quicksets grown, or permanently if further mounding were impracticable.[43] In some places, balks long continued as boundaries between fields in severalty, as mounds between field-course shifts, and as divisions between different parts of a single close whose ownership or tenure was divided, as when part was freehold land and part copyhold.[44]

Another practice, found in common and severalty fields alike, was the tillage of terraces along the contours of hillsides. In the first instance, the platforms were likely made in the same way as hillside roads were, by ploughing downhill and with some assistance from picks and shovels. Greensward cliffs were left between the terraces. In ploughing along the contours, as was usual in such locations, the furrow-slices could not be turned uphill. In some districts the cultivators used a turnwrest plough that could throw the slice either to the right or to the left. Elsewhere an ordinary plough was first driven one way to throw a slice downhill and then jacked up to return idle ready to turn the next slice the same way as the first. Either way the soil was always turned downhill.[45] We have two good descriptions of such terraces. Cobbett says,

I saw, on my way through the down countries, hundreds of acres of ploughed land in shelves. What I mean is, the side of a steep hill made into the shape of stairs, only the rising parts more sloping than those of stairs, and deeper in proportion. The side of the hill in its original form was too steep to be ploughed, or even to be worked with a spade. The earth, as soon as moved, would have rolled down the hill; besides the rains would soon have washed away all the surface earth and have left nothing for plants of any sort to grow in. Therefore the sides of hills, where the land was sufficiently good, and where it was wanted for the growing of corn, were thus made into a sort of steps or shelves, and the horizontal parts . . . ploughed and sowed, as they generally are, indeed, to this day.

And William Marshall writes,

The artificial surface that meets the eye, in different parts of these hills, forcibly arrests the attention. It occurs on the steeper slopes; which are formed into stages, or platforms, with grassy steeps, provincially 'linchets', between them. This form of surface must have been produced, at great expence, in the first instance, or by great length of time, in constantly turning the furrows downward of the slope. But as the turnwrest plough has never, perhaps, had a footing, on this division of the chalk hills, it is probable, that the stages under notice were formed, by hand; at some period when manual labour, either through an excess of population, or through the means of feudal services, was easily obtained. And the advantages, arising from the operation, have no doubt repaid the first cost, with ample interest. The stages, or platforms, are equally commodious for implements of tillage, as for carriages; beside retaining moisture, better than sloping surfaces; while the grassy steeps, between the arable stages, afford no inconsiderable supply of herbage; on which horses are teddered, or tended, while corn is in the ground; and which gives pasturage to sheep at other seasons. This sort of artificial surface is common, in different parts of the island; and the antiquary might be less profitably employed, than in tracing its origin.[46]

Terraces of this kind, with their intervening 'linches', 'linchets', 'linchards' or 'walls' were to be found in the tenth century at Tichborne, in the seventeenth at Amesbury,[47] and were widely distributed hereabouts in what we call the Chalk Country, as well as in the Chiltern, Southdown, Cotswold, Northwold, North, Peak-Forest, West and Poor Soils countries.[48] When they went out of cultivation in various places is imperfectly known; but at Clothall in the Chiltern Country, and in parts of the Isle of Portland, platforms wide enough to take modern implements were still being cultivated in recent times.[49]

Quite apart from balks of these various kinds, the common fields were far from being composed merely of tillage ridges. They also included whole ridges left to sward over and called, in the Midland Plain and thereabouts, 'leys', as opposed to tillage 'lands'. Many headlands, too, were laid down as 'headleys'. Commonly, whole ridges were laid to grass, but sometimes part of a ridge was tilled and part left in grass. And there were often found whole sets of ridges, and even complete furlongs and fields, in large leys suitable for common meadows, cowpastures or milking greens. Common permanent grass of this kind formed about half, and an integral and essential part, of early modern common fields in the Midland Plain[50] and the Western Waterlands.[51] Similar permanent leys were found in the Vale of Evesham,[52] the Cheshire Cheese Country,[53] and other plain and vale countries.[54]

In the least fertile soils in the down, wold and heath countries, common-field land was often laid to grass in order to recuperate from exhaustive tillage, and here, too, the resulting permanent leys were used as common meadow or pasture. Thus, in the Cotswold Country, in early modern times, the common flocks and herds depended largely on these greenswards and ploughing them up was either forbidden altogether or allowed only by general consent or if as much land were laid down elsewhere. Those who ploughed up their ley lands without proportionately reducing the number of sheep they commoned, were liable to prosecution. But the leys were generally, as they were intended to be, permanent grassland, so permanent, indeed, that they were sometimes eventually invaded by undergrowth and bushes.[55] In the early modern Oxford Heights Country, sheep were often fed on field land laid to grass, not merely in single ridges but also in whole furlongs and fields. One reason for this was that some of the soils were too light and thin to bear fallow stirrings and had perforce to be left in long still fallows to gain a sward.[56] Thus it is ordered at Bremhill in 1579 that 'noe tenante shall have goinge upon the layne sande fyldes but three shepe for every acre that he ther hath and two shepe for every acre that he hath in the vallowe clay fieldes.'[57] Chalk Country common fields likewise had some ridges laid to grass to be fed when the field was stubble or fallow and mown for hay when it was cropped.[58] Here, too, it was usually forbidden to plough up the leys, for they were needed to feed the common flocks.[59] And, as elsewhere, whole common fields were sometimes laid to grass. In 1574 a survey of Burbage records that 'the East Sandes hath byn a common feeld and arrable, but nowe used for common of pasture' throughout the year.[60] Laying to grass was also expedient when, as at Broad Hinton in 1636, there was too much tillage and too little pasture and meadow.[61] The barrenest tillage might then be selected for laying down. Unless manure were abundant and readily accessible, as in the immediate vicinity of sheepcotes,[62] the lighter and thinner soils could soon become barren and had to be rested from time to time and left untilled for two or three years to gain some sward.[63] Thus, in 1595, although 'usuallie eared', eighty acres of Collingbourne Kingston demesne farm had 'layne still uneared for the barraynenes of the same, and yet arrable land.'[64] At Manydown in 1650 one farmer had thirty-five acres of 'barren arable' with some herbage growing in it.[65] Northwold fields were similar to Chalk Country ones in that they commonly needed to be left in still fallows for several years.[66] And the same applies to other chalky countries. What were called 'laynes' in the Chalk Country and thereabouts, were termed 'ollands' (old lands) in the Norfolk Heathlands and the Breckland, but they were all much the same.[67]

CHAPTER 2

The origins of common fields

Common fields originated in early and remote times. They are first recorded in the Dooms of King Ine of Wessex, who reigned from AD 688 to 726. One of his laws reads, in part,

Gif ceorlas gærstun hæbben gemænne othe other gedalland to tynanne and hæbben sume getyned hiora dæl sume næbben and etten hiora gemænan æceras othe gærs gan tha thonne the thæt geat agan and gebete tham othrum the hiora dæl getynedne hæbben thone æwerdlan the thær gedon sie. – 'if churls have common meadow or other deal-land to fence and some have fenced their deal, some never, and their common plough-acres or grass be eaten, go they then that own that gap and make amends to them others that have fenced their deal for damage that there be done'.

Judging from later practices, the responsibility for dead hedges or fences defending corn and hay fields was laid on commoners in one of two ways: each man had either to fence any of his lands lying on the edge of the field, or to put up or mend his share of the whole fence. The passage quoted is worded in such a way as to refer to either of these procedures. As well as its most general sense of an open space, the word 'field' had several agricultural meanings: either a topographical area, or a field-course season, or an area under a particular crop. Fencing was not always put up all around a town's fields, for some adjacent towns intercommoned, nor all around topographical fields, for these were often bounded by rivers or other natural features, nor even all around seasons, for these were often bare-fallowed; but it was essential around each field of crops. Each crop of grain, pulse or hay had to be defended against wild and domestic animals. This meant each crop had to be defended individually. Thus the tenants of Fovant had to make up their hedges for the wheat field before Martinmas, for fallow crops in the hook field before 25 March, and for the barley field before 3 May. At other times and places we find separate special fencing orders for bean, pea, tilth and hay fields. It follows that each and every cultivator could reckon on having to fence part of his land at one time or another. So, whichever way one looks at it, Ine's law is proof positive of common fields and common meads (though not of permanent tillage) in Wessex in his time.[68]

'Gedalland' (deal-land) consisted of deals of arable or meadow, but excluded common pasture, which remained undivided. 'Gedalland' or 'dalland', we insist, means not partible land, but land that has been parted or divided and dealt out in shares. Bosworth-Toller errs in rendering these words as 'partible' land, though Toller's *Supplement* defines 'dalmæd' as 'meadow held in common and divided into doles or shares among the holders'. Partibility is beside the point, for all land is partible to owners who want to part it, and partible inheritance is not involved; inheritance is not as much as mentioned. 'Gedal' means division or deal, and 'gedalland' is land that is actually being dealt out or has been dealt out already; it is the 'deal-land' of early modern English, where 'courtland' (demesne) that is being or has been dealt out becomes 'courtdeal' and land dealt is 'deal-land'.[69] When Bishop Wilfrid grants 'sumne dæl londes' at Clifford Chambers, he means certain deal-lands, not just nondescript pieces of land.[70] 'Thes landes in under eal IX and XX gedale' signifies that the land Peterborough Abbey held in Maxey consisted of twenty-nine deals.[71] A corresponding verb was also used. Two brothers are granted 'twegra landa to gedale', that is two lots of land to divide or deal out between them, not merely to share between them, for it is possible to share something that remains undivided.[72] Likewise, when Archbishop Oswald grants 'VII æcras mædue on thæm homme the gebyrath into Tidbrihtingctune feorthe halfne on anum stede and feorthe halfne an othrum stede alswa hit to gedale gebyrath', he intends the seven acres of meadow in Tibberton to be granted as it should be dealt out, that is equitably, with half in one place and half in another.[73]

The testimony of Old English charters is no less precise than that of Ine's doom. Deal-land is often recorded. In 974, at Cudley in St Martin's Without Worcester, lay 'XXX æcrea on thæm twæm feldan dallandes withutan ...' – 'thirty acres in those two fields of deal-land without ...'.[74] At Clifford Chambers, where we have already found 'sumne dæl londes' in 922, we meet also, in 986, 'other healf hid gedallandes and healf hid on thære ege' – 'other half hide of deal-land and the half hide on their edge'.[75] At Maxey, some time between 963 and 992, Peterborough Abbey's small estate could hardly have been other than one of several similar holdings each consisting of deal-land.[76] We are even permitted to see such deal-land coming into being, and, incidentally, in parcels of unequal size, when the religious community at Worcester allows Bishop Oswald to declare,

dæt ic moste gebocian twa hida landes on Mortune on threora monna dæg minum twam getreowum mannum Beorhnæge and Byrhstane twæm gebrothrum and se eldra hæbbe tha threo æceras and se iungra thone feorthan ge inne ge utter swa to tham lande gebyrige. – 'that I may book two hides of land in Moreton on three lives to my two trusty men B. and B., two brothers, and the elder to have the three acres and the younger the fourth one, both inner and outer, as belongs to that land'.[77]

Similarly in 966, this selfsame St Oswald leased to one Eadric, amongst other lands, 'æt Uferan Strætforda on thære gesyndredan hide thone otherne æcer and æt Fachanleage thone thriddon æcer feldlandes' – ' at Upper Stratford in the divided hide every other acre and at Fachanleigh every third acre of field land'.[78] In both Upper Stratford and the unidentified town, the land has been, or is now being, dealt out. This kind of dealing has already been done at Bishopton when in 1016 are granted,

XV mæd aceras on thære ea furlunga forne gean Tidingtun et nigothe healf æcer on Scothomme and XII aceras yrthlandes betwyx thare ea and thare dic æt thæm stangedelfe ... and thone thriddan æcer bean landes on Biscopes dune – '15 acres mead there in Waterfurlong over against Tiddington, and the ninth half acre in Scotham, and 12 acres of earthland [tillage] betwixt the river and the ditch at the stonequarry, and every third acre of bean land on Bishops Down'[79]

When field land was dealt out in this way, the deals became intermixed in common fields. Thus at Charlton, near Wantage, in 982, we find some tenements *cuiusdam loco sed communis terre*, so that *rus namque praetaxatum manifestis undique terminis minus dividitur quia iugera altrinsecus copulata adiacent*[80] – 'in another place but of common land, so that the aforementioned land can hardly be distinguished by clear bounds on all sides because the coupled day-works on the other side border it'. In other words, the situation was the same as at Harwell in 985, where the lands were *segetibus mixtis* – 'in intermixed cornfields'.[81] The usual English expression for such intermeddled lands was simply 'acre under acre'. Thus at Kingston Bagpuize, at some time between 975 and 978, we find the following descriptions of bounds: 'This sind tha landgemæro to Cyngestune æcer onder æcer' – 'These are the landmeres to Kingston acre under acre' – and 'This sind tha landgemæro æcer under æcere'.[82] The same is expressed even more explicitly for Hendred in 962 (unless we side against Stenton with Drögereit, who declared the charter spurious): 'Thises landgemæra syn gemæne sua thæt lith æfre æcer under æcer' – 'These landmeres be common as they lie ever acre under acre'.[83] Similarly, at Drayton in 958, 'This sind tha landgemæra to Draitune æcer under æcer' accords with the statement in 983 that both Drayton and Sutton were *in communi terra*.[84] Like significance attaches to the 'gemænre mearce' – 'common meres' – at Winterbourne Bassett in about 972.[85]

Not all common land was necessarily common-field land in the sense of common tillage, but the existence of common meadow suggests that of common tillage also, for common meadow was normally dealt to men in proportion to their holdings of common arable.[86] It would therefore appear reasonable to infer that the common land recorded in all the following instances was common field and largely tillage. At Curridge in 953, 'on than gemanan lande gebyrath tharto fif and sixti æccera' – 'in the common lands belong thereto five and sixty acres'.[87] At Aston, in Dumbleton lordship, in 995 and 1002 respectively, two grants specify holdings *in communi terra* and *sorte communes populari*.[88] A charter in 997 describes an estate in Westwood, and then goes on, 'thonnæ licgeath tha threo gyrda on othære hæalfæ Fromæ æt Fæarnlæagæ on gæmænum landæ' – 'then lie the three yards on other side the Frome at Farleigh in common land'.[89] One cannot envisage sixty-five acres or three yardlands in common land, at those times and places, that were not largely tillage, especially when 'land' is used in contradistinction from 'pasture' and 'meadow'. Similarly the charter of about 972 relating to Winterbourne Bassett describes 'fif hida be Eastan tune gemænes landes on gemænre mearce' – 'five hides by East town of common land in common meres'. These hides almost certainly included tillage between the Winterbourne and Hackpen Hill, in East as distinct from West town, which correspond to the two ends of the township later called East field and West field, both of them much tilled. Incidentally, anyone doubting the essential continuity of rural England from the tenth through to the sixteenth century, or harbouring suspicions as to what was meant by 'gemæne', should look at this charter again. It specifies 'fif hida land gemæra into Winterburnam be Westan tune syndries landes ... fif hida be Eastan tune gemænes landes on gemænre mearce' – 'five hides of land mered in Winterbourne by the West town of severalty land ... five hides by East town of common land in common meres'. Here we see already the characteristic feature of Chalk Country towns in the sixteenth century and after: they were divided clearly and sharply between severalty fields and common ones, and between 'sundermed' and 'gemænen meade' – 'severalty mead' and 'common mead'. 'Syndries' and 'syndrig' mean precisely 'severalty' and 'in severalty', as distinct from 'common'.[90]

Now we come to what some might consider the clearest of all the records of Anglo-Saxon common fields, in that they give their evidence in the most readily recognisable form. In 961 at Ardington, 'Thas nigon hida licggeath on gemang othran gedallande feld læs gemane and mæda gemane and yrthland gemæne' – 'These nine hides lie in among other deal-lands, common field-leas and common meads and common tillage'.[91] Then in Avon in 963 three tenements lay *singulis iugeribus mixtim in commune rure huc illucque dispersis* – 'in single day-works promiscuously dispersed here and there in common field'. And at Upthrope, in 869, we are explicitly told, 'ægther ge etelond ge eyrthlond ge eac wudoland all hit is gemæne' – 'equally pastureland, ploughland, eke woodland, all it is common'.[92]

Next we advert to the *Rectitudiness Singularum Personarum*, or Rights and Duties of All Persons, a conspectus of estate management drawn up in about the year 1000. Under the heading 'Be Hægwearde' we find the following statement:

> Hægwerde gebyreth thæt man his geswinces lean gecnawe on tham endum the to etenlæse licgan fortham he mæig wenan gyf he thæt ær forgymth thæt him man hwilces landsticces geann thæt sceal beon mid folcrihte nyhst etenlæse fortham gyf he for slæwthe his hlafordes forgymth ne bith his agnum wel geborgen. – 'To the hayward it belongeth that one reward his swink in the end that lieth to the pasture-leas, for that he may expect if he first neglect this, which stitch of land be given him shall be by folkright nighst the pasture-leas, for that, if he by sloth his lord neglect, nor be his own well defended.'

By itself, this can hardly be taken as proof of common fields. Assuming, as the document suggests, that the hayward was the lord's own servant, answerable only to him, it follows, of course, that he was to be rewarded with a stitch of demesne land, very likely in severalty, for it would have been the demesne land he was expected to defend. (Usually, another or town hayward was elected by the homage to defend their own crops.) Nevertheless, the document has already told us not only that the beadle was likewise to be allowed a stitch of land, but also that cotsettlers should have five acres each and the 'gebur' or boor a yard of land, of which seven acres were to be sown by the previous occupier before the holding was made over.[93] With all these different holdings in fields where two or three-field courses were followed, it is reasonable to infer that common fields existed.

Where, then, the Chalk, Oxford Heights, Northwold, Cheese and Fen countries, the Breckland, the Vale of Evesham and the Midland Plain had evolved as distinct farming countries by the sixteenth century, permanent common fields already existed in the tenth. Obviously, they had not all been created at a stroke; but that such features as acres and headlands were used to demark bounds in Anglo-Saxon charters shows that the permanent arable fields had often already been extended to more or less their early modern limits. One of the relevant charters points to the current creation of new and further common fields, for what else could be implied by the grant to two brothers, when the elder was to have every first acres and the younger the fourth? Other charters in the tenth and later centuries, concerning Bishopton, Normanton-on-Trent, Brandon, and Hotham, grant or lease every second, third, eighth or ninth acre in the fields, suggesting, by their regular distribution of parcels amongst small numbers of men, either brand new or recently formed common fields. But Glastonbury Abbey courts of recognition in 1189 already show full common fields essentially indistinguishable from those in the Chalk, Cotswold, Cheese and Butter countries in the sixteenth century.[94]

In fine, the evidence plainly shows that the introduction of common fields in (but not throughout) England started before 726 and had run much or most of its course by about the year 1000. We conclude, moreover, that these common fields, like the severalty ones, were, by the tenth century, mostly in permanent, not shifting and temporary, cultivation. Otherwise it would have been both untrue and pointless to have specified the states of their land. Both the standing customs and the charters making grants of land for long terms habitually distinguish between earthland (tillage), meadow, pasture, and woodland. This alone suffices to prove that cultivation was largely permanent, for where the tillage plots are wholly temporary and shift from place to place, the land has no permanent state and cannot be identified by describing it either as tillage or as pasture, or even as woodland, it being now one thing and now another. And permanent tillage could only have been managed in field-courses with fallows, whose existence is proved, as soon as court records become available in the thirteenth century, by references to hitching or hook fields given over to fallow crops.[95]

We have already alluded to the sale in the late ninth century of *omnem octvam acram in Brandune* – 'every eighth acre in Brandon' – in the Breckland, and have taken this to mean that common fields either existed there already or were in process of being created. Of the existence of common fields in other parts of East Anglia before the Conquest there is a good amount of inconclusive evidence. We have wills dating from in and about the eleventh century that mention the acre and the yardland, 'aker', 'æcer', 'gird', 'gyrd', and so on. The yard of land, which was a quarter or similar fraction of a hide, and the acre, which was a day's work with the plough, were both intimately connected with tillage, and when they appear in a context that seems to imply the division of permanent tillage between a number of proprietors or occupiers, they suggest the presence or creation of common fields. Enclosed fields were not usually described in acres even when in tillage, both because square measurement was still impracticable or grossly uneconomic and because acres were not yet units of square measure. Acres were still, and were long to be, nothing but tangible field parcels, no two equal in size, and some remarkably bigger than others. Only if a field, or, for that matter, a meadow or a wood, were held, or to be held, in common, or if it were to be divided up, would it readily spring to mind to describe a parcel or piece of it as one or more acres. It would be more natural to say, for instance, one close by this or that name. Yet we find bequests cast in terms simply of acres, as the one to Somerleyton church of 'sixtene eker londes and enne eker med', with no further description, for this was to say so many acres in the one field or one set of fields in that town.[96] This field, or set of fields, must have been either in severalty or in common at the time of the bequest; if in common, then the common field was continued by the grant; if in severalty, then the grant would have led the acres to be distributed between the sown and the fallow lands, and perhaps even to be intermeddled in various parts of the field or fields. Such developments would explain why charters after 1086 relating to the heathlands of Norfolk make mention of 'delelond', 'deleacres' and 'dels'.[97] Taken together, all this evidence suggests the early presence of common fields not only in the Breckland, but also in precisely those parts of East Anglia where we know of a certainty they existed in later times, to wit, in what became the Norfolk Heathlands, East Norfolk and High Suffolk.

Sparse fragments of a once larger body of evidence compel us to believe there were common fields in England in the eighth century and permanently cultivated ones in the tenth. To argue, as has been done, that common fields were absent in the years anterior to those for which records of the manorial regulation of field-courses are extant, is frivolous. Halmote records, if ever made, have not come down to us from Anglo-Saxon times. That common-field regulations fail to appear in extant manorial court rolls, early or late, may be due to a variety of reasons.

First, in small common fields such matters were usually settled by neighbourly agreement and needed no sanction from manorial courts. Why detailed regulations were not needed in court ordinances in the North and Peak-Forest countries is explained by the ordinances themselves. They say, for instance, that 'where anye man is plaintif in neyborheade' he is 'to come to the counstables and cause them to geve warnynge to the neighbors to come togeather and se the same reformed.' Most agreements for small fields could be made and disagreements settled merely 'by neighbourhood', without formal presentments, orders, trials and penalties in manor courts. As late as 1663, common pasture stints were not always recorded in the court rolls, implementation being left to the neighbours themselves.[98] The management of common fields by neighbourhood was usual in most of the Northdown Country. In the Chiltern Country, the contrast between large and small fields, in this matter, stands out clearly. We know that rights of common were regulated and stinted, and that there were common flocks and herds, with common herdsmen and haywards paid by rates assessed on the commoners and levied by virtue of the sanction conferred by manorial courts.[99] But whereas in the great common fields along the escarpment, field-courses were laid down and enforced by the manor courts, they were elsewhere decided by informal and usually unrecorded agreements between the occupiers of adjacent lands.[100] Partly for similar reasons, court rolls tell us little or nothing about the field-courses in the Lancashire Plain and the Cheshire Cheese Country, though they record many orders relating to ditches, strays, encroachments and common pastures. In 1612, for example, the Leyland freeholders and their tenants drew up a full agreement for the regulation of the town fields, including the depasture of sheep and cattle and the tethering of cattle and horses on ley lands 'when the said feildes go abroad and are pastured'. Boundary balks were to be fed off or ploughed up only by agreement between the neighbours concerned. The numbers of cows, sheep and horses to be put into the field were stinted and unregistered agistments forbidden. But as for field-courses and cropping, no word was set down about either.[101]

Secondly, there were probably some general but unrecorded customs, just as there were unwritten common laws. For instance, it was a general custom or law that each proprietor (or his tenant) should do his share of fencing off crops in the common field. Thirdly, not all general orders, even if clearly enunciated, were necessarily deemed sufficiently important to warrant promulgation in the manor court, and therefore could not be recorded in the court rolls. Fourthly, sub-manors had their by-laws recorded in the rolls of the head manor and groups of manors sometimes had theirs promulgated and recorded in their hundred courts.[102] Fifthly, promulgation and enrolment were not always carried out methodically. Just as enrolled by-laws were occasionally not enforced, some enforced ones remained unenrolled. No field-course regulations were recorded in the Weyhill rolls, not because there were no such regulations, but because there was no need to have them recorded.[103] Even when by-laws were promulgated, the townfolk might spare themselves the expense of enrolment. This we know from the special requests made to court stewards that certain orders be engrossed and enrolled. Unless so requested, the steward had no need to order enrolment, and the clerk would write nothing.[104] After all, the court crier had to have his fee and the steward made a charge for every entry of presentment in the court roll, as well as for signing and taxing every bill of costs, and though these fees were modest, yet the pressure to keep down local rates was perhaps no less in those days than in ours.[105] It would normally have been a waste of money to have had the custom for the fencing of fields enrolled. It is significant that even the full and comprehensive early modern manor court rolls rarely record details of common-field courses or of fencing regulations.

Sixthly, it was hardly practicable to record by-laws not previously drafted in writing. The usual procedure was for the steward, upon request, to instruct the clerk of the court to enrol certain by-laws as drafted, amended and approved. What a clerk liked best was to be handed a bill, perhaps as amended in debate, of the laws to be recorded; otherwise he would have had to have taken down, in long hand, every word as it was promulgated in open court. Furnished with such notes by the town officers, the clerk would then transcribe the orders before filing them away with the rest of the court documents. It was from this sheaf of drafts and rough notes and his own minutes that the clerk had to engross and enrol the by-laws and write up the whole court roll. But all this demanded more literacy and latinity than was commonly found among Anglo-Saxon rustics. The progress of the enrolment of common-field regulations may mark the advance of literacy and of written records, but is no guide to the advent, rise and spread of common fields.[106]

It has also been argued that early common fields, because of the vast area of rough grazing available, had no use for common of pasture in the herbage of the stubbles and fallows. But the animals would have preferred this to rough grazing, and their depasture needed regulation. Moreover, fallows, even still ones, had to be either mucked or folded. As soon a fire without fuel as a common field without either farmyard manure or the sheepfold. To do any good, hundreds of sheep had to be close-folded on the tillage, so either the lords (or perhaps their tenants independently) set up large joint flocks and folds, which would have necessitated detailed regulation, or each individual occupier owned hundreds of sheep, which he, by himself, family and servants, would have daily driven to feed on the rough grazings by day and brought back to fold on the tillage at night. He would also have had all the lambing and shearing to attend to. All this would have preoccupied him to such an extent as to leave him little time for growing cereals. If sheep were folded in the early common fields, the folds would have been joint ones. If they were not folded, the land must have been mucked with farmyard manure, from which it follows that either the field must have been tiny, or the individual herds and flocks must have been yarded or housed in enormous farmyards, sheds and cotes. In the absence of evidence, it is impossible to see exactly what was done in the early common fields; but as soon as evidence becomes available it proves beyond question that in permanent common fields, while some small and easily accessible parts were mucked with farmyard manure, most of the tillage was usually close-folded by joint flocks according to strict regulations.

Sparse and scanty as the Anglo-Saxon records be, they prove that permanent common fields existed. But they tell us little about them and fail to reveal their dates of origin. And though they suggest what caused their genesis, they afford little evidence of this. Man's frailties being what they are, in the absence of clear evidence, the human mind, thus liberated from historical discipline, and its imaginative powers given full play, is all too apt to plunge into an orgy of unbridled speculation. And so it is here. Indeed, the fantasies surrounding the origins of common fields are almost unrivalled in all the annals of mankind.

Seebohm proposed that the dispersal and intermingling of properties in common fields originated in what he called co-aration, by which he meant the use of composite, joint plough-teams formed by the beasts of two or more men for the purpose of communal ploughing. Acting communally in this way, he supposed that the peasants who had jointly tilled the land then allotted it in equitable shares and ridge by ridge amongst themselves, so creating common fields.[107] Unfortunately for this theory, no evidence has ever been found of this kind of co-aration in England other than on demesne lands the lords kept strictly in their own hands, where no field could be common. Practised on tenantry lands, such co-aration could conceivably have led to common fields, but it is precisely here that evidence of co-aration is totally absent.[108] Some early modern probate inventories itemise half a team, but only along with half a plough, half a cart, half a cow, half a crop, and so on, showing either a business partnership, as when sons inherited and kept the holding intact, or an as yet undistributed estate when the deceased had willed an equal division of his goods.[109] For the rest, we are left merely with informal arrangements by which two neighbours lent beasts or teams to each other from time to time, as when Nicholas Walker and William Ellyot of Broadway were 'joyning their teemes togither in ploweinge'.[110]

Co-aration was, indeed, practised among Irish and Welsh extended families; but this could not have created common fields, for, contrary to the tenor of the forged document Seebohm relied on, it was not followed by the allotting of land into individual properties, and a field in which there are no separate and distinct properties cannot be in common.[111]

A widespread belief in theories based on co-aration has been somewhat sustained by misapprehensions about the significance of town ploughs. Gutters, trenches and drains in common fields were often made by a common or publicly owned drainage plough provided for by a special rate levied for this express purpose and drawn by a great composite team made up of eight or nine yoke of oxen led by one or two pairs of horses, or of up to a score or so of horses. The husbandmen held their plough or drove the team they had contributed to; the plough-boys rode on the guide-horses; and the labourers followed behind with spades and shovels to perfect the work.[112] To drain their stiff, miry, clay fields and pastures, the common-field farmers of Caxton used a trenching plough with a mould-board three times the normal length, two coulters and a flat share, all provided and maintained at the public charge. Drawn by twenty horses, jointly supplied, this plough could cut a trench one foot deep, a foot wide at the bottom and eighteen inches at the top.[113] At Naseby, the town or common plough, likewise furnished with two coulters, but drawn by a composite team of no more than ten or a dozen horses, could cut a drain one foot wide and deep and throw out the earth in a giant furrow-slice on the right hand.[114] This was rather like the co-aration envisaged by Seebohm, except it was not concerned with tilling the soil and could have done nothing to institute common fields in the way he supposed. Co-aration is but an imagined cause of common fields. Even if co-aration had ever taken place in the way Seebohm supposed, there would still be no reason to link it with the creation of common rather than severalty fields. If two or more owners or occupiers of land in severalty had formed a joint team for tilling the soil, they would still have had no compelling reason to set up a common field between them. The evidence is non-existent and the reasoning illogical.

Orwin, however, gave a new twist to Seebohm's explanation by arguing that it did not necessarily depend on co-aration by composite plough-teams:

When the ploughing season began, there would be the lord's team, the teams of the owners of hides, and the composite team of owners of virgates and bovates, all of them at work. If they went to work daily side by side, the result would be an automatic alternation of strips between the plough-owners, those contributing to the composite teams having proportionately fewer strips than the owners of whole teams ... The men of the community would go to work, setting out a day's ploughing for each team, side by side. By the end of the working day each would have ploughed one or more 'lands', according to the nature of the soil and the length of the 'lands'. Next day the village husbandmen would move on to set out and then to plough the section of the field next beyond the first day's work, and thus they would proceed throughout the ploughing season.[115]

This is pure fiction. How could the lord's ploughs, drawn as they were by his tenants' beasts, go to work at the same time as their ploughs did? And even if the selfsame beasts had been in two different places at the same time, why should all the teams have worked side by side? Unless their portions had already been allotted, how would the ploughmen have known how to space themselves out? We would need to know why the land was allotted as it was in order to know why it was set out as it was. Anyway, no evidence is shown, so historians need pay no heed.

Let us instead commence our enquiry with the evidence, meagre as it is, that relates to the institution of common fields. Strange as it may seem, we have some evidence of the creation of common fields in the early modern period. The first class of common field whose origin fortunately remains recorded is that of temporary common fields in shifting cultivation. Dispersion arising from one great division was the rule in temporary common fields, 'in ridge by ridge', 'rivings', 'stitchmeal', and 'lot acres', where arable was allotted in direct proportion to the common rights enjoyed in the rough grazings broken up. Proprietors and tenants met and arranged the allotment of ridges in a fixed order of dispersal, one man having, say, every sixteenth ridge, so that his land was scattered throughout the field. Each man's particular ordinal number was decided by lot, just as in common meadows everywhere and just as in the land distributed *sorte communes populari* in Dumbleton in the year 1002.[116] We are told exactly how this was often done: 'So many lots are put into a hat and everyone is to take his lot as it happens', preferably getting an illiterate to draw out.[117] An allottee's ordinal number was up to chance, but the regular dispersal of parcels minimised any disparities. Chance and certainty were happily combined in equity. We also have evidence concerning the creation of new permanent common fields in early modern times. In the Chalk Country, the farmers of West Grimstead seem to have carved a fourth field out of the down,[118] and shortly before 1700 the tenants of Avon were licensed to plough up land 'which was highway, and now called New Field, and is to be continued arable and sowed in course as a fourth field'.[119] At much the same time a similar development was seen in Ibthroppe,[120] and apparently in Knighton also.[121] It is hard to see how the men of West Grimstead or Avon could have managed without casting lots, spinning coins, or some play of chance. Anyway, one grand division seems to have initiated dispersal and intermingling. Similarly, in some enclosures of common fields in this farming country in the eighteenth and nineteenth centuries, not all the common rights in them were extinguished; some allotments were laid out as new common fields and common meadows, entailing the setting out of dispersed parcels. Four such new common fields were instituted at Fovant, four at Stoke Farthing, four in each of the north and south tithings of Broad Chalke, and similar ones at Urchfont.[122] The measured shares were scattered and distributed throughout the fields, partly because each allottee needed his land more or less equally divided between the four shifts for the four-field courses adopted. An additional and powerful reason for dispersal was put forward in one of the awards:

> And whereas it is by the said Act provided that, for the accommodation of various proprietors, lessees and other persons ... who are desirous of having the same allotted into larger parcels and more convenient situations as aforesaid but by reason of the smallness of the allotments they would be entitled unto and the very considerable expense of keeping a man to attend the depasturing of sheep or other cattle on such small and unfenced allotments in soils and situations where the difficulty of raising substantial quick fences is almost insurmountable, such proprietors also being desirous of having liberty to feed and depasture their allotments in common ...

In short, they wanted common flocks and common folds under common shepherds in common fields, and arrangements were made accordingly. This meant their parcels had to be intermingled, for how else could the common fold be moved equitably? Only by intermingling could each man be assured he would be neither the first nor the last to enjoy the fold passing over his land. In the words of an eye-witness, 'As the necessity of a common sheep flock still continues for the sake of manuring the common-fields lands, a considerable part of these small properties are still occupied in their original state of commonage.' Common folds of one kind or another were used in most common fields, and among the motives for dividing them into dispersed parcels was that of ensuring the equitable distribution of soils and locations, of lands within shifts, and of the benefit of the fold.[123]

In addition, and elsewhere, in hill-farming regions like Wales and the Peak-Forest and West countries, 'butty' fields held 'in common' by two or three men were formed by the sharing either of old closes or of new intakes from the moors and forests. But such butty fields were often subsequently divided into separate closes.[124] 'Communal activity tending to the enlargement of the common fields' has been supposed in much earlier times, but the examples Bishop gives relate to districts where shifting cultivation prevailed, and where permanent tillage, let alone permanent common field, was very small in extent. Sowerby and Bramley, like Rochdale, lay in what became the Peak-Forest Country; Sproxton, Everley, Easby, Kirkby Knowle, and Ingleby Greenhow in the Blackmoors; and Bramhope in the North Country. Leaving aside a few doubtful locations, the only other places concerned are Burton Leonard, Ribston, and Bramham Bigging, all straddling the bounds of the North Country and the Midland Plain, and partaking of the North Country practice regarding intakes. This body of evidence seems to relate solely to the formation of temporary common fields in shifting cultivation. What is most striking is that the vales of York and Stockton, the northern extension of the Midland Plain, where permanent common fields were the general rule, were not found to yield instances of their augmentation by such assarts and intakes. Authentic examples of the extension of permanent common fields by the subsequent division of intakes originally in severalty are far from common.[125] Nevertheless, the township had an important role in the extension of common fields. Assarts or intakes from stinted common pastures, marshes or woods always had to be allotted to all the men with common rights in those places and to each man in strict and direct proportion to his rights of common there, these rights in turn being in proportion to his existing holding of land. So when it was the general desire to extend common fields at the expense of stinted pastures, the township as a whole had to agree to it, as at Harlestone when the heath was broken up.[126]

Furthermore, old-established closes were often split up amongst co-heirs or other beneficiaries, by custom of descent, by will, by bargain and sale or by lease. But whether it be a new intake or an old close that we find divided makes little difference in principle, for the old close was itself once a new intake. What often emerged from such divisions was a number of small enclosed common fields. However, let us be wary of exaggerating the facility with which such common fields could be created. It was not divided ownership that led to common fields, but divided occupation. An assart divided between two or more owners or tenants could nevertheless be cultivated in severalty by one single lessee or sub-tenant. By the same token, an intake owned entirely by one man could be divided among several petty tenants and cultivated by them in common.

It is evident that all common fields arose from the division of land previously not cultivated in common and that this original division was the sole characteristic shared by them all. But the mere division of land between two or more men cannot of itself create common fields. The division of pastures often led to a multitude of small closes, and the same enclosure awards that divided and allotted land in new common fields, also divided common fields and allotted them in severalty and in enclosures. Whether the allotments were made in common or in severalty was a matter for conscious choice.

The way in which divisions were made is of some consequence. In existing common fields or meadows, or in the creation of new ones, and in land reclamation generally, the usual method of division in earlier times was by splitting each tract of land lengthways. In an existing tillage field, this meant lengthways of the furlongs and ridges, and in an intended common field, temporary or permanent, it meant splitting the intended furlong lengthways of the intended ridges, usually by measuring off along its ends with a chain, cord, rod or yardstick.[127] Similar means were adopted in event of the consolidation or rectification of existing parcels.[128] Before the middle of the sixteenth century and the invention of the plain table and the theodolite, accurate and economical square measurement was not generally practicable. Square measurement is still too costly today to employ in the division of fields for building plots and allotment gardens. It has always been cheaper and easier to measure along the two ends and split the whole into long parcels.[129] Splitting arable in this way was conducive to the formation of common fields, and common-field parcels were frequently divided and subdivided in the same way, on inheritance, sale or lease, often until they were too narrow to be split again, unless, perhaps, for cultivation with the spade. The width of the parcels was thus clearly a function of the degree of division and subdivision.[130]

But the process could be reversed, and side by side with the increased morcellisation of some parcels, especially but not exclusively in later times, went the consolidation of others. The frequently piecemeal nature of such consolidation is well exemplified by the history of the Nicholas family of Roundway as they slowly and laboriously exchanged their parcels in the common fields with other holders and regrouped them into larger pieces.[131] The eighteenth century, however, saw several instances of planned and general consolidation, similar to that now financed by the European Community in western Germany and other parts of the Continent.[132] Yet it would be wrong to conclude that there was no such consolidation in earlier times, or that all large parcels were amalgamations of smaller ones, or that all demesnes in common-field townships were at one time in common field. There are many examples of demesnes being extricated from common fields and put into severalty, and some of the regrouping of lands when lordships were divided;[133] but signs are not wanting that some demesnes had always been in severalty, even in otherwise common-field townships.[134]

When division and intermingling had progressed to a certain extent, agreements and regulations were needed to determine when the whole furlong or field should be fallowed, when fenced against livestock, when open to them, and so on. Not surprisingly, the earliest court rolls record intermingling and infractions of such by-laws, witnessing yet again that the common fields already had a long history behind them.[135]

Here, then, is an explanation of the way common fields developed in permanent cultivation. But this explanation will not do for temporary common fields; they lasted for too short a time to allow any considerable degree of subdivision. Fortunately, for temporary common fields an explanation of origins has been readily forthcoming from the evidence and needs only explication. We know that the tillage was allotted in direct proportion to the rights of stinted common pasture forgone when a plot of rough grazing was selected for the plough. If this plot were virgin soil, each commoner had his share measured out by the yardstick. Then the next time this plot was brought into tillage no such measurement was required, for the pattern of ridge and furrow was still on the ground and allotment could proceed simply ridge by ridge. The reason such plots were held in common was that one commoner by himself could not prepare a plot for tillage. In temporary common fields in all the hill countries, it was essential to muster all the sheep, cattle, horses, and goats belonging to all the townsfolk in order to make up the running fold that alone would suffice to fertilise the soil well enough to allow crops to be taken. This fold had to move from one piece of the plot to another until the whole had been covered. The only convenient way of organising this was to design a plot of contiguous parcels, and the only equitable way, lest one man should be able to broadcast all his seed well before another could even make a start with his, was to have these individual parcels intermixed.[136] Thus no less in temporary than in permanent common fields, the common fold was the agent compelling the common management of intermixed and intermingled parcels. In permanent common fields a small area near the farmsteads could be mucked with farmyard manure, but this was too scanty to serve for the whole field and too difficult to convey to its more distant parts. Temporary common fields were sited in rough grazings too far away from the farmsteads for farmyard manure to be carried to them. Even when cattle were being summered up in the hills, farmyard manure was not produced there, and the only practical way of fertilising the plot was by means of a running fold.

At first sight, then, we appear to have two separate and distinct but equally compelling reasons for the formation of common fields: yardstick measurement, and common folds. But on closer inspection these two causes are revealed as deriving from one common cause at one step removed. Division of the land between a number of men gave rise to yardstick measurement and allotment; and the effectual division of the flocks and herds between many small proprietors gave rise to separate parcels of sheep and cattle too small for folding unless combined into common flocks and herds. The land had to be folded, and the division of the township's land and livestock among a number of small proprietors necessitated common fields and common folds. We now have a comprehensive historical explanation of the causes both of the origin and of the continuance and persistence of common fields.

This explanation will be found to fit all instances and occurrences in all parts of England at all times, and goes to show an ounce of evidence is worth more than a ton of speculation. Even such exceptions as we shall find will serve to prove the rule. In the Vales of

Hereford, where sheep were not folded in early modern times, a main pillar of common-field husbandry was missing, and the fields themselves vanished all the sooner. In countries where farm property was not much divided and subdivided, in High Suffolk, for instance, common fields never attained the sway they did elsewhere and were the sooner enclosed. In the western extremes of the Northdown Country, some common fields existed in much their pristine state throughout the early modern period. Here were found stinted commonings of sheep and beasts in fields and downs, regulations for husbandry and some enforcement of field-courses.[137] In the main body of this farming country, however, common fields were mostly minute at this time, partly because great tracts of open land, it seems, had always been in severalty, and partly because former common fields had been enclosed or otherwise put into severalty. Since the common fields were mostly small and were not town fields common to all the township, field-courses needed no ordering in by-laws and were left to informal neighbourly agreement, while the fields themselves could all the more easily be engrossed by a single occupier and common rights extinguished by unity of possession.[138] By the early modern period, there were no common flocks or folds and no apparent distinction in cultivation between common, severalty and enclosed fields. Common fields survived only nominally or in small fragments, and common downs continued, even in name, on only a few chalky swells.[139] Even where large common fields had originally been cultivated, they disintegrated in earlier times, as at Wye,[140] apparently by the same kind of piecemeal extinction of common rights that was still being carried out in the extreme west of the country in the early modern period.[141] Why this happened so easily may be guessed from the husbandry practised on the chalky soils. The deeper and loamier of them, originally in permanent tillage, were so fertile that year-long bare fallows were soon replaced by fallow crops of 'podware', so making common rights less advantageous. The sheep-pastures were well suited to tillage, but only in a system of convertible husbandry, like that followed at Westerham in the Middle Ages.[142] Ploughing up this land took the ground from under the feet of the common flocks and made common-field practices redundant. The dissolution of common flocks and folds eliminated the part-time and family farmers who depended on them, and once these had gone, there was no point in having common fields.[143] As, also, the demesne lands, though occasionally dealt out to small tenants who cultivated them in common, were mostly in severalty even in unenclosed districts, common fields were hardly to be found in the early modern period.[144]

It was, then, the division of land and livestock among small proprietors, accompanied by lineal measurement and fold-mucking, that gave rise to common fields and ensured their continuance. By the same token, it was the accumulation of property and stock into fewer hands that entrained the abolition of common fields.

But we have still not got to the bottom of the matter, for we have yet to explain why and how land and livestock came to be dispersed into many hands. One obvious occasion for division was inheritance by co-heirs, each of whom would have received an equal share following the splitting of each and every parcel of land and the equal division of the livestock. Another, and more frequent and widespread, occasion was the alienation of part of a holding by bargain and sale. In practice, we may suppose, these two kinds of transfer fitted nicely together, with some purchasers dividing property between heirs and some heirs selling their inheritances. Such sequences of events led to some whole entities of land being split into common fields and common meadows with common of pasture, estover and turbary, and to the further subdivision of these common lands and rights. This explanation was first proffered for parts of Wales by Owen in 1603 and was repeated by Davies in 1814. Gray showed that some small common fields came into existence partly through the division and subdivision of closes between heirs and assignees. In the Chiltern Country, some small arable closes, when split, became common to two or more men. This clearly must be regarded as one of the ways, very likely the main one, in which small common fields were created both in Wales and elsewhere. Yet it was not the custom of partible inheritance that caused the partitions, but the desire for partition that led to the adoption of the custom. When repeated partition led to disadvantageously small holdings, the custom itself was overridden by disgavelling statutes and charters.[145] Moreover, the creation of common fields cannot be deduced simply from the existence of a custom of partible inheritance. It is necessary to show evidence not only of a custom of descent by partible inheritance, but also that custom was not defeated, as all customs of descent could be, by willing and devising some other course of descent, and if not, that the divisions upon inheritance actually created common fields. It can never suffice to show the custom and then to assume the creation. One must prove the custom operated to produce the result. This it would only do in the absence of a will to the contrary. Then one would have to prove that two or more of the heirs took up their shares of land and that one heir did not buy out the others. Even when all the heirs took up their shares, it was still open to them to work them all in partnership, and then no common rights could have resulted, so not only division into shares, but also the independent occupation of shares, has to be shown.

For inheritance and assignment alone and unaided to have caused the rapid division and thorough dispersal of lands into common fields, they would have to have been accompanied by an increase in the density of rural population. Conversely, a fall in this density would have been accompanied by the consolidation of common-field parcels. But events of this kind can rarely be traced, because we have no exact knowledge of the size of the population at various times. It seems there was a rise in the population in the twelfth and thirteenth centuries, but we frankly guess there was a stronger upsurge from the seventh to the ninth century or thereabouts.

We have, however, no call to be disheartened because we cannot know what the population was at various times, for neither do we know that it was an upsurge in population, in conjunction with partible inheritance, that gave rise to common fields. Indeed we believe such a theory to be wholly fallacious. Division gave rise to common fields and perhaps some of this division was accompanied by rural impopulation, but we have no reason to suppose that any increase in rural population was essential to the formation of common fields. When a landholder, who has been employing two of his sons, dies and leaves the land to be divided between them, this division results from a decrease in population, not from an increase. This is admittedly an extreme case, but it alerts us to the possibility that the birth of common fields was not immediately caused by a rise in population. Indeed, the causal relationship may have been the other way round: we must open our minds to the possibility that it was the institution and progress of permanent cultivation and of common fields that caused a rise in the population.

We readily grant that repeated partition between heirs and assignees sometimes gave rise to common fields, albeit mostly small ones, and that it led also to the minute subdivision of parcels in established fields. But likely some more powerful force operated in the same direction. If we assume that, anterior to partition between the heirs of conjugal families, agricultural improvements permitted, at some stage, the division of a relatively small number of old families into a greatly increased number of smaller ones, then a radical and rapid division of lands and livestock and the sudden rise of common fields would become distinct possibilities. As the land was tamed and its fertility and productivity increased sufficiently to allow shifting and temporary cultivation to be replaced by permanent cultivation with permanent tillage in which crops and fallows were alternated, a great increase in crops could have caused the population to rise rapidly through earlier marriage. At first this change may have come about slowly, on a limited scale, and almost imperceptibly; but we may hazard a guess that at some point it would have rapidly speeded up and attained momentous proportions.

In the first clearing of scrub and woodland for agriculture, shifting and temporary cultivation, often of the slash and burn variety, would have been practised in England, just as it still is today in various parts of the world. But on their first cultivation in new clearings, brown woodland soils, which are created by leaf fall, deteriorate rapidly, so shifting cultivation has willy-nilly to be practised until such time as the land can be regularly fertilised with farmyard manure or the sheepfold.[146] Now, when they abandoned their tribal organisation, in western Europe in prehistoric times, the Germanic peoples are generally assumed to have settled in hamlets each consisting of a single patriarchal family, with several generations living, working and sharing all in common. The patriarchal family, or something approximating to it, was well suited to pioneering, clearing woodland, breaking up virgin soil, taming the wilderness, and engaging in shifting and temporary cultivation. But the innovation of permanent cultivation changed everything. Now a premium was placed on the loving attention to detail that came best from a conjugal family working a small holding. So the patriarchal family gave way to a number of conjugal ones, each of which then severally enjoyed its own allotted share of the tillage, livestock, implements, and common rights in the rough grazings, fallows, stubbles, woodlands, meadows and marshes. So, in place of a single patriarchal family constituting a primitive township, there emerged a number of conjugal families living as neighbours in a common-field township.[147]

A yet more powerful force seems to have been the effective cause of the creation of systems of large and regular common fields. If we concentrate our gaze on the familiar landmarks of the recorded history of this period, we may recognise four salient features: cohesion, conversion, manumission, and the genesis of common fields. Now, let us suppose that instead of a patriarchal family breaking up into conjugal families, or of a father dividing his lands and goods between two or three sons, we find a lord dividing half an estate or plantation between one or two score slaves or colonists. Would this not be a force potent enough to create townships of large, regular common fields?

The Romans owned slaves and so did the Anglo-Saxons. Slave-owners and slaves changed, but slavery as an institution continued. If our ancestors of the early ninth century were in anything more advanced than Americans in the United States in the early nineteenth, it was not because they had no slaves, nor even for that they more frequently manumitted them, but for that they were more prone to hut them and to divide large parts of their plantations between them.

Long before the fall of the Roman Empire, slavery was a failing institution. As few slaves could be bred in captivity and there was a limit to the number of criminals to be punished by penal servitude, the chief source of slaves in ancient times was in war and conquest. Men, women and children were taken prisoner and enslaved. When they were conquering the 'barbarians', the Roman aristocrats had plenty of slaves. But eventually the legions took to the defensive, and slaves, being fewer, became ever more costly to buy and use until their labour became uneconomic. Owners solved their problems by hutting their slaves, giving them allotments of land, allowing them to raise families, and making them fend for themselves, while still exacting from them forced labour, now only part-time, on those parts of their estates they kept in their own hands. In this way both sides were advantaged: the one had some freedom and more opportunity and the other had a self-perpetuating supply of superior labour. With the irruption of the 'barbarians' and the foundering of the Empire came a reverse wave of invaders, conquerors and enslavers. Slavery got a new lease of life in the British Isles, as chief fought against chief, king against king, and English against Welsh. But eventually a united English kingdom was formed and internecine strife came to an end. As the sources of slaves therefore dried up, so slaves were hutted and even manumitted. In

Anglo-Saxon times, indeed, manumission and hutting often went hand in hand, as when a lord willed that 'all my men are to be free, and each is to have his homestead, and his cow and his corn for food'. Such actions were close to the spirit of Christianity and conversion impeded further enslavement, for one Christian might not enslave another. The great divide, however, was not between slaves and freed men, but between a plantation of slaves and a lordship over men who, even when they were still legally slaves, were hutted and in charge of their own holdings and economies.[148]

Side by side with the hutting of slaves went similar divisions and allotments to colonists in newly founded lordships. Furthermore, by one means or another, men formerly free became bound to lords and incorporated into lordships. Thus a great mass of villeins was created by the depression of the free as well as by the freeing of the *servi*, serfs, slaves or bondmen. Irrespective of whether the tenement or *mansus* was *mansus colonica* or *mansus servilis*, of whether the *cassati* or hutments were inhabited by men who were legally slaves, all tenements set and let by the lord owed labour on his demesne lands, or corn to his granaries, or money to his coffers, or some combination of two or more of these or similar services; service-tenancy had arrived.[149] Now, we know this allotment of lands was roughly contemporaneous with the rise of common fields. All the indications point to economic and social change rather than population growth having been the grand cause of common fields, and nowhere is this more apparent than in the earliest surviving evidence of these fields. The various charters are concerned with grants of five *cassati* at Curridge, ten *mansi* at Drayton, nine *mansi* at Ardington, three *mansi* at Alveston and Upper Stratford, three *cassati* at Hendred, three *cassati* at Avon, three hides at Bredicote, two at Moreton, two *mansi* at Clifford Chambers, ten *mansi* at Winterbourne Bassett, one *mansus* at Cudley, seven and thirteen *mansi* at Kingston Bagpuize, seventeen *cassati* at Harwell, five *cassati* at Charlton, and two and twenty-four *mansi* at Dumbleton. As for Ine's Doom, it speaks of 'gedalland', which is dealt land. Deal-land and common field were synonymous terms in the seventh century, as they still were in the seventeenth. Significantly, the selfsame documents that record the common fields also record *cassati* and *mansi*, hutments and hutted tenements and messuages carved out of the estate (or domain in the wider sense) and dealt out to tenants of the lordship.[150] The tenantry and their common fields are recorded jointly because their origins were joint. Fields were dealt out to tenants under such circumstances and in such a way that they could not but be common.[151] In much the same way, in later times, the division of demesne lands, hitherto in severalty, amongst the tenants of a manor, extended old, or created new, common fields.[152] (That common fields were occasionally newly dealt out, as a result of previously intercommoning townships agreeing to go their own separate ways, or of the extrication of demesnes from common fields, or of the division of a manor between two or more lords, merely indicates the strong hold common fields already had.)[153]

We are thus faced with three main sets of circumstances under which common fields might arise: the creation of service-tenancies, the change from patriarchal to conjugal families, and inheritance by co-heirs coupled with sale and purchase.

Among the English, in early historical times, patriarchal families were unknown; in the records, only conjugal ones are met with. This suggests that the break-up of the patriarchal family and the initiation of common fields had occurred before the eighth century. Such hides and other large holdings as continued as single agricultural entities could not be managed unaided by individual conjugal families. The churl with a hide or so of land had to rely on assistance from his slaves, underlings and dependants. But, as the techniques of permanent cultivation were further diffused, holdings of this kind gave way to those worked by service-tenants, by geburs with yardlands and cotsettlers with a few acres each. But patriarchal families, spanning three or four generations and with all their property common to the kindred, survived among the native Irish until the nineteenth century, among the Scottish highlanders until the eighteenth, and among the Welsh until the fourteenth.[154] Much of the division and subdivision of lands amongst these peoples, in the past so often attributed to simple inheritance by co-heirs, seems to have been caused mainly by the dissolution of patriarchal families. In sharp contrast, distribution to service-tenants seems to have been most marked in the more central and western parts of England. And it is impossible not be struck by a roughly conterminous contrast between the great permanent common fields of the plains, lowlands, downlands and wolds on the one hand and the usually small or temporary common fields of the mountainous districts on the other. It seems we should think in terms of separate histories in distinct geographical districts: the history of common fields in the regions where the cultivation of demesne arable originally required a great labour force of service-tenants, and the history of common fields in predominantly pastoral regions. And, then, in addition, we must consider local variations in soils, terrains and climates.

It is tempting to try to explain the different origins and characteristics of the various field systems along racial lines. The difficulty is the apparent lack of exact geographical coincidence between clearly distinct racial regions and recorded variations in agrarian institutions and practices. Yet there is no denying that the English occupied the parts where great permanent common fields emerged and left the western mountainous regions to the Welsh. Gray's suggestion that no permanent fields obtained in the regions of so-called 'Celtic' culture was not wildly wide of the mark. Few such fields have ever been found other than in places where the English settled, i.e. in England, the lowlands of Scotland, and the Englishry in Wales, and these few may safely be attributed to later imitation.[155]

We need to look again at the early history of ploughs. The Romans had no ploughs with mould-boards to invert furrow-slices. The assertion that Pliny the Elder (AD 23 – 79) said such a plough had recently been invented has no foundation. What he actually wrote was,

non pridem inventum in Raetiae Galliae ut duas adderent tali rotulas, quod genus vocant plaumorati; cuspis effigem palae habet. serunt ita non nisi culta terra et fere nova [*novali* ?]*: latitudo vomeris caespites versat, semen protinus iniciunt cratesque dentatas supertrahunt nec sarienda sunt hoc modo sata, sed protelis binis ternisque sic arant*[156] [or, in a restored and better reading, ... *quod genus vocant ploum Raeti* . . .].[157] – 'Not long ago an invention was made in Raetia by fitting a plough of this sort with two small wheels, which the people call a Raetian plough; the share has the shape of a spade. They sow this way, but only on cultivated, and usually new [fallow?] land. The breadth of the share turns the turves over. Men at once broadcast the seed and draw toothed harrows over the furrows. Nor by this method need the sowings be hoed, but they plough after this fashion with teams of two and three yokes.'

Pliny is describing a paring plough with a skim-share and with wheels to ensure its even pitch and uniform depth of working. It was designed to pare the turf off plots being brought into temporary cultivation.[158] No one riding on the Rhaetische Bahn southward from Landquart through the Graubünden (Grisons) could doubt the utility of such implements of supplement the ordinary *aratra* used there and in the Alpine region generally.[159]

The Romans came to have many kinds of *aratra* or drag-ploughs far removed from the primitive implements associated with earlier agriculture. In addition to the Raetian ploughs, there were ones with iron coulters and shares, and sometimes with wheels, ones with shares with 'ears', wings or flanges for ridging, and heavier ones better able to deal with stiff or virgin soil. But none of these was intended for, or suited to, inverting furrow-slices and ploughing ridge-and-furrow fashion. The Raetian plough turned over slices of turf, but not furrow-slices. Some ploughs made ridges to deepen the seed-bed, but they inverted no furrow-slices. They had no mould-boards.[160]

The early history of ploughs has been bedevilled by the misapprehension that the chief distinction to be made was between heavy wheeled ploughs and light ones without wheels. Yet wheeled *aratra* have been widely used in places as far apart as Slovenia and Sweden,[161] while many heavy swing ploughs, complete with mould-boards to invert furrow-slices, have had, instead of wheels, a foot to slide over the wet soil. The point about wheeled mould-board ploughs is not that they have wheels but that they have one small or land-wheel to run on the land about to be ploughed and a large wheel to run in the furrow that has just been cut. The crucial distinction to be made is between ploughs that invert a furrow-slice and those that do not.

That no Belgic or Romano-British mould-board has ever been found proves little, for early mould-boards were fashioned in wood and have perished. Plough-irons have been found, but no number of coulters and shares can tell us whether or not the wooden frames they came from had mould-boards attached.[162] Nevertheless, it is generally agreed that the Romano-British coulters found on Twyford Down, at Chedworth and elsewhere, are most unlikely to have been fitted to mould-board ploughs.[163] The discovery of iron shares, including winged ones, proves nothing, for many *aratra* had them. Even the unearthing of an asymmetrically winged share in Dinorben fails to prove it was used on a mould-board plough.[164] First, it is uncertain whether the asymmetry was by design or due to wear and tear. Secondly, a share with a single wing was an obvious choice for use with *aratra* on hillsides, in order to work the soil downwards. Thirdly, just because a plough had a share in no way proves it had a coulter. Peaty and fenny soils have often been cultivated with a plough fitted with a one-winged share and an earth-board, but no coulter. Such a plough can make a furrow and push it to one side. But there is a world of difference between scuffling up the soil and moving it to one side, and cutting, inverting and throwing a furrow-slice.[165] What these various finds of coulters and shares really prove is that the Belgae and Romano-British used some much improved types of *aratra*.[166]

Small squarish fields bounded with earthen banks an open hills and heaths are sometimes called 'Celtic',[167] but were not peculiarly so. The bounding of fields with earthen banks is not a practice peculiar to any race or age; it is a common expedient where walling, fencing and hedging are all impracticable.[168] On the high chalk downs, for instance, many such fields seem to belong to the Iron Age, but some were cultivated in and after the Middle Ages. Such fields were not thrown into ridge-and-furrow, but ploughed, and often cross-ploughed, with a drag-plough or cultivator.[169] That these 'Celtic' fields were mostly squarish in no way argues that fields in similar locations but less square were formed by mould-board ploughs. The long fields dubiously attributed to the Ancient Britons prove nothing about the way they were ploughed, for such fields could be, and often have been, ploughed and formed by drag-ploughs or *aratra*.[170]

Finally, it can hardly be purely fortuitous that all these fields, coulters and shares have been found in thin soils on downs, hills, and wolds. At least as far back as Defoe's days,[171] it has been generally understood that more remains of antiquity are to be found in these relatively undisturbed parts than elsewhere. Nevertheless, that they are found where they are is itself of some significance. It stretches credulity to breaking point, for instance, to be asked to believe that a mould-board plough was used to cut and invert furrow-slices at Dinorben hill-fort on the barren Denbigh Moors.[172] In fine, the Ancient Britons confined their tillage almost entirely to the more easily worked soils and used nothing more advanced than improved or specialised *aratra*.[173]

Yet this is not to say that the Romans and Ancient Britons refrained altogether from ploughing heavy soils in the vales and plains, only that they could not have cut and inverted furrow-slices and provided surface drainage.[174] And although the Ancient British engaged almost entirely in shifting cultivation, this cannot be regarded as peculiarly Celtic,[175] for temporary common fields, and shifting cultivations in general, obtained in early modern England, in the North-eastern Lowlands, the Breckland, the Norfolk Heathlands, and the Sandlings, Chalk, Southdown, North and Peak-Forest countries, and in the poor sandy soils in the Midland Plain.[176]

The plough that for the first time allowed the cutting, inversion and throwing of furrow-slices was fitted with a coulter, a one-winged share and a mould-board. Any implement, wheeled or not, that could be fitted in this way the English called a 'syl', 'sule', 'sulh', 'solh', 'suluh', 'suolu' or some such, and later, 'sull' or 'sullow'. In the Danelaw, the word 'plough' came to be applied to the sull also. In Wessex, however, 'plough' continued to mean not the implement, but the team that drew a sullow or a wain. Since a sull might be either wheeled or footed, the word *carruca* is inapposite. *Carruca* means any wheel-carriage or any cart or plough with wheels.[177]

The origins of the sull are unclear. No coulter suitable for a sull has been found in England dating from before the eighth century.[178] Eleventh-century illustrations of what seem to be wheeled sulls may conceivably have been copied from ninth-century Continental ones and, anyway, may be regarded by some as insufficiently detailed or accurate for technical analysis. The artists seem to show no mould-board, but they clearly depict a plough with a combined sheath and sole or ground-rest, to which a mould-board could easily have been fitted when required,[179] and this seems to have been exactly what the word 'sull' meant. No written description of a sull has been found as early as this, but the word 'sull' is found in ninth-, tenth- and eleventh-century English and the English seem to have brought it over with them.[180] The claim that the Saxons who dwelt between the Elbe and the Oder already possessed wheeled ploughs with mould-boards to turn furrow-slices and were employing them in the Elbe marshes in the first century AD, before they came over, is not fully warranted by the evidence. The ploughs then used in the marshlands of Lower Saxony seem to have been heavy wheeled implements, but no evidence has been found that they had either coulters or mould-boards or could either cut or invert furrow-slices.[181] They apparently had fixed earth-boards that pushed the soil to one side, but no coulter to cut furrow-slices. The implement used in these parts in ancient times was in all probability some version of the *Beetpflug*, which had a pair of wheels of equal size, a fixed, flat, rectangular earth-board, and a one-winged share, but no coulter. It made low ridges and shallow furrows and was best used in long fields or furlongs, but could not cut a furrow-slice and was unable to provide surface drainage in heavy soils.[182] Nevertheless, fragmentary and frail as the evidence is, it is difficult to escape the widely accepted conclusion that the sull and the ridge-and-furrow system of ploughing were invented by the Germanic peoples for the purpose of tilling the moist and heavy soils of north-western Europe and that these same inventions were introduced in England by the English during the Dark Ages.

Since the ridge-and-furrow system of cultivation was one of the foundations of English husbandry and was intimately associated with almost all the permanent common fields created in England, it is futile to try to explain the differences between various types of these fields by reference to peoples other than the English.

All attempts to explain the distinctive features of common fields in East Anglia by supposed Danish or Friesian influences have failed. The division of the lordship of a town between two or more manors, though more frequent there, was not peculiar to East Anglia. Nor was the exchange of parcels by means of subtenancies and conveyances in which the parcels were described by the names of the tenements they had originally belonged to, for this practice was not found in all parts of East Anglia and yet obtained later in parts of the Vales of Hereford. The *tenementum* was not an institution peculiar to East Anglia as a whole, for it was not found in the fenlands there. Nevertheless, in the greater part of the old kingdoms of East Anglia and Kent, tenements were notionally preserved as means of identification and as units of assessment for taxes and services,[183] signifying, it seems, that common fields in these parts arose less from the planned and orderly distribution of lands amongst service-tenants and more from the haphazard division of large family holdings into smaller ones. And this would seem to accord with what we know of the manner in which the Germanic peoples first settled the land in those places. It seems to have been more by infiltration and less by conquest than it was in Wessex and the Midlands.[184] This circumstance, and probably others created by the mode of settlement, may have affected both the emergence of common fields and their internal arrangement and management, but not the methods of cultivation, which differed radically between the various regions of each kingdom, but in each were marked by the widespread cultivation of leguminous crops.[185]

That the town fields of the Cheshire Cheese Country and the

Lancashire Plain were small and covered but a small part of the farmland, and then often on the lighter soils, was not due to Celtic influence, if any, but to grass being a more suitable and profitable crop than corn. As large sheepflocks could not be kept for folding the fallows, only small acreages of permanent tillage were feasible, and as the common fields had no purpose beyond ensuring domestic supplies of bread and drink corn, there was no point in enlarging them. If wheat was not much grown, this was due to agricultural conditions, not to racial characteristics. If the field-course in town fields in the Lancashire Plain only allowed for spring crops, this was nothing to do with race; it was because the climate was unfavourable to winter corn.[186]

That there were only small areas of common field in the Northdown Country and the Wealden Vales, the High Weald and Romney Marsh, is explained better by the wet and heavy soils and dense woodlands than by fiscal *iuga* (yokes) or Jutish settlement.[187] The ascription of holdings to heirs and children in rentals[188] was not necessarily due to the implementation of a custom of partible inheritance, for it was common form among estate agents in recording a property whose actual ownership was unknown because rentals and scutage rolls had not been kept up-to-date.[189] Moreover, while some fields were made common by successions of sales and purchases or of divisions between heirs or by the leasing out of demesne lands in small parcels, there were anciently common fields in the towns on the flanks of the downs where the Jutes first settled.[190]

Gray's work was invaluable because it demolished the view then current that common fields with two- and three-field courses formerly prevailed throughout the kingdom. He argued that distinctive racial characteristics had an influence on economic development. There is much to be said for his view that 'The nature of field systems depends primarily upon the relation of the unit of villein tenure to the arable fields. For this reason it is pertinent to inquire in what measure the systems ... are Anglo-Saxon, Celtic or Roman'.[191] It is possible to find permanent common fields in what modern politicians have dubbed 'Wales', but this bloated administrative monstrosity includes large tracts of England and of the Englishry in Wales occupied, settled and cultivated by English people. It is a striking fact that all the ancient permanent common arable fields in Wales were in the Englishry, in parts settled by Englishmen and Flemings, for example, in South Pembrokeshire's 'Little England Beyond

Wales', Gower and the Kidwelly and Lackarn districts. Once the English had invented suitable field-courses for permanent tillage, they introduced them in the former Welsh lands they occupied. How great a change they wrought thereby may be seen from their activities in the vale of Clwyd, around Denbigh. The Welsh had been unable properly to cultivate what were some of the richest and deepest soils in the kingdom. The English replaced temporary by permanent cultivation and laid out planned and regular common fields.[192] There is good reason to suppose that the forms of permanent cultivation suited to conditions in England, and to some other parts of north-western Europe, were invented in England by the English. Here racial characteristics were not merely evident; they were of crucial importance.

But racial differences, so significant in the invention of this form of permanent cultivation, had no influence on the incidence of the various field-courses practised in permanent tillage. As Gray says,

> Two and three-field methods of tillage were not expressive of racial or tribal predilection ... What determined the adoption of one or the other form of tillage was agricultural convenience, and this in turn depended largely upon the locality and the nature of the soil ... Hence a change from one to the other was a matter of opportunism ... between these two modes of husbandry the difference was not one of principle but of proportion.[193]

Similarly, all the incomprehensibly tortuous arguments to the contrary notwithstanding, the small common fields of the Northdown and Chiltern countries are, Gray admits, best explained by reference to farming conditions in countryside 'so influenced in its field system by its topography that its original affiliations cannot readily be discovered.'[194]

In fine, common fields arose from division and allotment accompanying the creation of service-tenancies, the dissolution of large family holdings, and inheritance, sale and purchase. It remains, true, however, that common fields were allotted in ways that fitted in with the field-courses, with the lie of the land, the nature of the soils, inclination to the sun, and many other purely local circumstances.[195] It remains true likewise that men joining together in ploughing facilitated the cultivation of butty fields and the extension of common fields, though, obviously, it was the existing division and dispersal of property that necessitated such joint efforts. It is true, too, that common fields existed for the sake of the common of shack, for the division of land was inescapably accompanied by the creation of diminutive flocks and herds that could only be well managed in common. It is even true that town drainage ploughs played a part in the further development of existing common fields. Nor have we any reason to deny that the genesis and rise of common fields was accompanied by an increase in population. But what cannot be accepted is that a rise in population was the cause of the invention of common fields. Indeed, we would be on firmer ground if we asserted it was the hutting of slaves, the planting of colonists, and the distribution to them of lands in common fields that allowed them to multiply their own kind as never before. It would be absurd to regard customs of partible inheritance as a major cause of common fields generally. Where common fields were supreme, in Wessex for instance, there was little customary inheritance of land and even less customary partible inheritance. Conversely, where there was a general and uniform custom of partible inheritance, as in Kent and Wales before disgavelling, common fields were the exception rather than the rule. Nor could joining together in ploughing cause the inception of common fields any more than of enclosed and several ones. Nor is it credible that common fields were due to assarting, for assarts can as well be in severalty as in common. It cannot even be taken that the need for a common flock and fold caused the inception of common fields. It would be truer to say that common fields, flocks and folds all arose at the same time from the same common cause. Petty proprietors and tenants could keep no more than petty flocks and herds and needed common fields. By the same token, common fields were brought to an end by the consolidation of parcels of land and of livestock into large blocks and flocks in the hands of capital farmers[196] who had no use for common regulations, common of pasture, common herds, common flocks and common folds. In short, common fields had their formal cause in the methods of tilling and dividing agricultural land, and their immediate, proximate, final and effective cause in the creation, in Anglo-Saxon times, of petty tenancies, properties and occupations.[197]

CHAPTER 3

Field-courses

At one time common fields had obtained in all the farming countries except, it seems, the Vale of Taunton Deane, the Wealden Vales, the High Weald, and Romney Marsh;[198] but late medieval and early modern High Suffolk, East Norfolk, the Woodland, and the Sandlings, Saltings and Northdown countries had only isolated patches of common field.[199] In all these countries the common fields had similar features arising from a shared development, especially, it seems, from their original settlement by free men and from their precocious agricultural and general economic progress. The few small High Suffolk ones were distinguished by the relative compactness of holdings, by the somewhat limited distribution and dispersal of parcels, and by common of shack on the stubbles being the only generally and clearly recorded exercise of common rights and duties. The Sandlings Country had few or no wide stretches of common field; instead, small fields and furlongs were interspersed with small, irregularly shaped and scattered 'pightles' or closes formed from single parcels or furlongs, suggesting that the piecemeal enclosure of common fields had occurred as rapidly in the past as it was proceeding in the present. Common of shack was exercised in some places, but little record remains of any close regulation of husbandry by town meetings. Extrication from the common fields was facilitated by the low degree of morcellisation (or the height of consolidation), it being quite usual for one man to occupy all or most of a whole furlong or field.[200] Common of shack was also the only generally recorded common right in the irregular but often large common fields in East Norfolk in the Middle Ages. Whole-year bare fallows recurred only once every half-a-dozen years, the third year of three-field courses otherwise being sown to peas, beans, vetches and other spring crops and the land cleaned mainly by half-year bare fallows. These common fields were mostly extinguished in the later Middle Ages in the course of further advances in agricultural techniques.[201]

In the Butter and Cheese countries, the Western Waterlands and the Vale of Berkeley, small common fields existed in many or most towns and tithings in the early modern period, but were being rapidly extinguished. The reason was that these countries were best suited to dairy-grazing, and once the necessity for assured self-sufficiency in corn had passed, common fields could safely and profitably be enclosed.[202]

In Wales, the Blackmoors, the North-western and North-eastern Lowlands, and the West, North and Peak-Forest countries, what permanent common fields there were, were almost all extremely small; shifting and temporary ones were generally larger.[203] Other than by extending its application to the north, modern historical research can hardly better William Marshall's generalisation:

> In the western extremes of the island, the common field system has never, perhaps, been adopted; has certainly never been prevalent, as in the more central parts of England. There a very different usage would seem to have been early established, and to have continued to the present time, – when lords of manors have the privilege of letting off the lands of common pastures, to be broken up for corn; the tenant being restricted to two crops, after which the land is thrown open again to pasturage.[204]

In the North-western Lowlands, despite some small permanent ones sown every year to spring crops, most common fields were merely shifting cultivations or 'rivings', i.e. plots of newly broken grassland. Holme Cultram copyholders, for example, were accustomed to plough up the greater part of Colt Park for three years at a time and then to leave it unsown as stinted common for six years. There were also other lands called 'acredales' or rivings that were divided into four rivings, each in turn sown for three years and then thrown open as common pasture for nine.[205]

In the North Country, in the wider and sunnier valleys and on the gentler and warmer slopes, as in parts of Wensleydale, Bishopdale, Coverdale, Wharfedale and the Hexham and Craven districts, there were, in the sixteenth and seventeenth centuries, and even in the eighteenth, some small common fields, lying in several topographical divisions, each with its furlongs and their dispersed parcels bounded by greensward balks,[206] and all ordered by manorial courts. Some of these fields were apparently in permanent cultivation, but common rights in them were being increasingly extinguished by agreement.[207] Elsewhere, the area of common field was occasionally extended. Thus, by 1524, three common fields had been enclosed from the moors at Halton Shields for the town newly founded there.[208] Throughout the early modern period, moreover, intakes or 'improvements' were being made from the waste, and many of these and other closes were divided and cultivated as common fields.[209] But permanent tillage was always the exception; most was purely temporary. The only wide expanse of arable was in the moors and here tillage was necessarily severely limited both in size and duration. In cultivating the moors, by which were meant all common pastures whatsoever, it was usual for proprietors and tenants to meet together and decide what land was to be brought into tillage and how divided between them, always in proportion to the extent of their permanent holdings and of their appurtenant common rights in the moors.[210]

Permanent common fields were equally rare in the West Country. Marshall thought the dispersal and intermingling of enclosures in the vale of Exeter possibly indicated their former existence,[211] and other circumstantial evidence suggests the same here and elsewhere, especially in the South Hams. Brixham and Leigh (near Pillaton) had both had common fields once, and those at Braunton survived into the twentieth century. Permanent common fields, however, were clearly the exception.[212] 'Some parte of this mannor', remarks Norden at Leigh, 'lieth in common feilds, which is hardly founde in any mannor of his Highnes in Cornewall.'[213] Temporary common fields were much more numerous and widely distributed than permanent ones, for most soils were too fleet to permit tillage to be continued for long. Shifting cultivations in the moorlands eventually created enormous expanses of ridge and furrow. The custom was for the manorial tenants to divide and plough up the commons for a crop of winter corn on narrow stitches, to be followed perhaps by oats, and then to lay the plots down again for rough grazing in common, shifting the stitches from hand to hand and from place to place so that no man had the same land above two years together.[214] Much of the evidence that might appear to relate to permanent common fields is in fact concerned at most with temporary ones. Stoke Climsland has a common heath or down of a hundred and fifty acres where the farmers 'doe enclose parte of the said downe and sowe it for one yeare and then throw it open againe'. It is presented in Farway manor court in 1629 that, 'divers of the tenants ... did about xlvj yeers since plow and putt in tillage all the ... common called Knowlehill', which was thrown open after two years for common of pasture and for the collection of ferns. In Whitford, two hundred acres of common moor, covered with heath, ling and furze, 'hath heretofore bene in tyllage as yt dothe playnlie appere by the ridge and furroughe'.[215] Common fields like this were not necessarily open; on the contrary, some were common closes; but, whether open or enclosed, were usually described as being 'in stitchmeal', as, e.g. 'le common downe close in stichemeale', meaning that the stitches or ridges were methodically dispersed and intermingled in them.[216] In fine, lands in stitchmeal were open or enclosed common fields of temporary tillage in shifting cultivations in which allotments were made stitch by stitch to each of the commoners according to their properties and common rights.

The typical common field in the North-eastern Lowlands was likewise temporary. Even though a three-field course generally prevailed, successive seasons being sown to winter and spring corn and then bare-fallowed for a year, the whole course was shifted from place to place.[217] At Cowpen in 1599,

> At the layenge forth of any decayed or wasted corne feilde and takinge in any new feildes of the common wastes in liewe thereof, everie tenannte was and is to have so much lande in everie new fielde or, according to the rente of everie tenanntes tenement, in such place and places as did befall everie of them by their lott; and so hath everie of the quenes tenanntes within the towne of Cowpen aforesaide, as well leassors, tennannts at will, as freeholders, contynewed the occupacion of all their arable lands by partinge by lott as aforesaide; and that after the layenge oute of everie wasted corne feilde within the feldes and territories of Cowpen aforesaide, everie so wasted and layde oute corne felde nowe is and ever was reputed and used as the quenes common wastes there are, until the same lately layde oute corne feildes or any of them be by general consente of neighbours taken in, parted and converted to arable lande or medowe again.[218]

Even with bare fallows every third year, crops so speedily exhausted the land that it had to be suffered to relapse, often furlong by furlong, into rough grazing, while a fresh field was taken into cultivation in a system of 'ridge by ridge', each man's land being sprinkled 'rigge by rigge to his nighbour accordyng to the old devysion of lands in this countrye' – 'by rigg and rigg ... as is the custom in every husband town' – with the object of ensuring equitable allotments of good and bad land.[219] The new ploughings, called 'rivings' in the North-western Lowlands, were known hereabouts as 'rifts'. Thus in 1567 each tenant of Guyzance had 'the xvjth rigge in every new rifte which is to be maid arable and which before was lee or pasture ground'; the demesne farmer at Newham took a share in any 'new rifte' made; and at Haswell in 1570 oats were growing 'in the reffte'.[220] This shifting cultivation explains the changes in Shilbottle common-field names, the confused nomenclature of arable and meadow grounds at Cockfield, an order forbidding further ploughing in Green Field in Hartley, the 'wheat raynes' in Acklington ox-pasture, the designation of common tillage plots by 'sides' and 'quarters' rather than by fields, and the many temporary divisions and enclosures of common land, which are all too easily mistaken for permanent enclosures of supposedly permanent common fields and pastures.[221] Habitual common agreements and manorial regulation, made doubly useful here on account of shifting cultivation, assisted also in the greatest reform in common-field husbandry, viz. the redivision of whole townships, the lands of one town being apportioned between two, three, or even four new towns or tithings. The great advantage in this was in the saving effected in the time spent in going back and forth, particularly to and fro the tillage plots, for these were usually sited far away from the town's farmhouses. In the years after 1560 many such subdivisions were made and the land redistributed ridge by ridge by lot under the supervision of special commissioners and arbitrators. Yet it occasionally happened that redivision had reluctantly to be abandoned, either for the lack of water supplies to the new towns planned or because extreme diversities of soil made equitable allotment excessively complicated and difficult. Attempted subdivisions sometimes gave rise to prolonged and rancorous disputes, resulting in the modification or abandonment of the scheme. Thus petty jealousies and misunderstandings led to Chatton township being divided only into two and not into four as originally intended. In yet further instances, it proved easier to take the town to the tillage and the whole settlement was simply moved to a more convenient location.[222] Removals like this probably account for the scattering apart of towns, halls, mills and churches apparent in many townships.

In parts of Wales and most of Scotland and Ireland, the common fields consisted of infields and outfields. The infields were usually winter-fallowed and mucked for barley (generally of the bere or four-rowed variety), then given a single ploughing for oats, and so to barley again, and so on in a perpetual round. Occasionally peas, beans or even summer wheat were grown. The general rule was that the land was sown to spring crops of one sort or another every year. The outfields were, to a greater or lesser extent, either tathed or folded with all the livestock summering on the hills or other commons, or pared and burned, or rib-ploughed, all to prepare temporary and shifting common fields that bore only inferior spring crops like grey oats or, in Ireland, potatoes. The ratio of infield to outfield varied considerably from place to place. In the more favoured parts, like the eastern lowlands of Scotland, the infields exceeded the outfields in area, but elsewhere the ratios were reversed, often to such a degree as to leave only tiny infields, in Wales often mere gardens. Among the few parts where most of the common arable was in permanent cultivation were the coastal plains of the Moray Firth; the Pale, Ulster and other parts of Ireland effectively settled or planted by the English race, including Scottish lowlanders; and the Pembrokeshire, Gower, Lakarn and Kidwelly district. As was the usual way everywhere, the outfields were distributed ridge by ridge whenever broken by the plough. In Scotland such fields were said to be in 'runrig', in Ireland, in 'rundale'. In the remoter parts of Scotland and Ireland even the infields were redistributed annually, biennially or triennially, as befitted temporary cultivations.[223]

Even in farming countries where nearly all common fields were permanent, some temporary ones were still to be found in modern times. Temporary common fields were made in the heathlands at Mapledurham in the Chiltern Country in 1720–1.[224] When the 'falls' of the Northwold Country were commonable, one at a time was selected for tillage by agreement between the various parties and then divided, usually by casting lots, into the parcels of what was a temporary common field.[225] In the Sandlings Country, Dedham had temporary common fields in the heath.[226] Even in the Midland Plain such practices were not unknown. In some exceptional localities, we are told, 'The tenants divide and plough up the commons, and then lay them down to become common again; and shift the open field from hand to hand in such a manner that no man has the same land two years together'. A township would enclose and allot to its members different parts of the waste in turn for two or three tillage crops.[227]

Seeing these practices, and especially the inclusion of three-field courses in shifting cultivations and the temporary abandonment of some field land in the Chalk Country and elsewhere, we may be forgiven for imagining the way in which temporary tillage became permanent: by gradual reductions in the areas or durations of abandonment.

In order to find out what field-courses were followed in permanent common fields, it is first necessary to rid oneself of the notion that they were necessarily reflected in field-names. Like fields in severalty, common fields consisted of varying numbers of topographical divisions or fields in a purely locative sense. Take, for example, common fields in the Chalk Country. Some townships, like Fordington in Dorchester, had their common arable 'all in one most spacious feild',[228] while others, especially in the sixteenth and seventeenth centuries, had theirs in two topographical fields,[229] more in three,[230] and yet others in four.[231] Many towns, too, had a large number of fields, especially where the soils were largely flinty clay loams, encouraging the growth of woodlands that broke up the landscape. Sometimes these fields were too numerous to permit accurate counting and it can only be said that there were one or two score of them, that their dispositions were occasionally changed, their names altered, and their numbers increased or diminished.[232] The topographical fields, however, bore only the loosest relationship to shift fields and field-courses. At Basingstoke, for example, there were about a dozen topographical fields, but some of these were grouped with others into shift fields that formed units for the rotation of field-courses. When Hackwood Field lay fallow, so did Mill Field.[233] In the Southdown Country the common-field tillage, mostly on the lower slopes of the downs, was sometimes in two or three topographical divisions; but whether few or numerous, the fields were grouped or apportioned into 'laines', shifts or seasons for field-courses. Thus at Prinsted in 1640 a survey clearly speaks of 'this layne', in the sense of a tilth or season, composed of various topographical fields. Rodmell had a 'wheat laine' and a 'barley laine', i.e. a winter and a spring corn field.[234] The demesne farm of Northease and Iford had 126 acres of 'erable lande in iij laynes in several felds'.[235] It is understandable, then, that parcels of open arable were commonly described according to which 'laine' rather than which field they were in.[236] In the Chiltern Country, large arable fields lay along the arable foreland and the escarpment, and temporary ones in some of the heaths; but elsewhere, since patches of open field and enclosure were scattered here and there, the common fields were usually small and numerous. Each farmer had his lands dispersed in only a fraction of the town's many topographical fields.[237] None of this had significance for field-courses, for the fields were grouped into seasons according to their soils and other characteristics.[238] The topographical fields of the Northwold Country were numerically various, but the towns often had one or two fields above the wold and one or two below, the field-courses being quite different up and down.[239] Some townships in the Oxford Heights Country also had small and numerous common fields that could not have coincided with field-courses.[240] In most Cotswold Country townships, division into two fields reflected the field-courses, but in others the presence of many small fields masked them.[241]

In the eastern countries, the position was similar. The common fields of the Norfolk Heathlands were divided into various numbers of topographical fields and furlongs.[242] Little distinction, however, was made between furlong (*stadium, quarantina, quartrona*) and field (*campus*) in surveys, terriers and field-books. Instead, the parcels were often located by reference to the 'precincts' they lay in, a precinct being an area within certain natural boundaries, such as highways and rivers. Often these precincts (and sometimes also the furlongs within them) were designated merely by ordinals. Usually the lands of any one farm were grouped together in only some precincts, and often in a single one, suggesting that the common fields themselves had been started originally by several families rather than by a single lord. Neither topographical fields nor precincts, but shifts, formed the basic units for field-courses.[243] Distinctions between 'field' and 'furlong' meant no more in the Breckland. A field-book of Brandon speaks of 'one feild called Parsonage Furlong', and so on. For the purpose of field-courses, the ridges, parcels, furlongs and fields were grouped into shifts or 'shift fields'.[244] Similar practice and terminology are found in East Norfolk.[245] In High Suffolk and the Sandlings Country, common-field parcels were often disposed in furlongs ('wents') and topographical fields, and occasionally in precincts; but, once again, the terms 'field' and 'furlong' were interchangeable and signified little other than the smallness of the tracts themselves.[246]

In the plain and vale countries, field-names are hardly better guides to field-courses. In the Midland Plain, topographical fields were sometimes numerous, but were then grouped into shifts for field-courses. Berwick-in-Elmet, for example, had five fields, but no five-field course.[247] When Brigstock had five fields, each was in a three-field course.[248] At Stanford-in-the-Vale-of-White-Horse, of six topographical fields, four formed one single shift.[249] In Laxton, two of the four fields formed one season.[250] Glapthorne and Cotterstock had a dozen fields grouped into three shifts for a three-field course.[251] Similar tripartite divisions of numerous fields were found in many other townships.[252] In 1594 three of the Kenho fields lay in the first shift and many in the second and third, while one field was split between two of the shifts.[253] Conversely, at Barrow-on-Soar and elsewhere, unusually large topographical fields were divided up into shift fields.[254] In this part of the world, if a shift were found to be too small or too large, land was simply transferred from one shift to another.[255] In the Fen Country, likewise, though often disposed in three, four, or more topographical divisions, the fields had long been grouped into shifts for field-courses.[256] Maxey has two fields in each of three shifts. Sixteenth-century Sawtry has five fields, but Fen Field 'soylith with Wood Filde and Wood Filde with this filde ... being both one soyle'. East and Middle fields are likewise in one shift, whereas Far Field is 'one hole soyle of itselfe'. In Fen Ditton in 1672 'the tilth feild or feilds ... lyeth in two feilds or places ... both make but one feild', while 'High Ditch Feild and the lands ... which by custome have been sowne as parte of that feild, lye in severall places'.[257] Turning now to the Vales of Hereford, it is evident that each tenant did not hold land in each of the small topographical fields so frequently found. The fields were grouped, sometimes according to hamlets, always according to shifts, so each tenant was concerned only to have his lands dispersed between the shifts, not between all the fields.[258] The often numerous fields of townships in the Vale of Evesham were, in early modern times, generally arranged in four shifts, quarters or seasons. Shift fields were used also in the Vale of Berkeley and the Cheese and Butter countries.[259] Quedgeley, for example, had a score of common fields, Stoke-under-Hamdon, fifteen, plus several common closes, and all arranged into shifts.[260] In the Western Waterlands, widespread division into two great topographical fields suggests that field-courses above and below the hill differed from each other as much as the soils did, so, as in the Northwold Country, the number of fields has a significance quite other than might have been supposed.[261]

Let us, then, set aside field-names and turn to field-courses. Gray used the circumstantial evidence given by the even distribution of tenants' parcels between the fields, but this is not conclusive evidence of the nature of field-courses, for parcels were evenly distributed according to soil and terrain as well as according to seasons. Thus a town with one field above and one below the hill would have equal distribution between the two, but each field would be under its own distinct course or courses, which would be indiscernible from the distribution of lands in the fields. Conclusive evidence of field-courses is thus to be found only in records of field-courses.

Field-courses in permanent common fields fell into three classes, which, for reasons soon to become apparent, we have named as follows: plain-country, downland and Lancashire Plain.

The ordinary plain-country course, practised in the Midland Plain,[262] and also in the Western Waterlands, the vales of Evesham,[263] Pickering[264] and Berkeley,[265] and the Butter,[266] Cheese,[267] Fen,[268] and Cheshire Cheese[269] countries, in 'Little England beyond Wales'[270] and in heavier soils in the Chiltern Country,[271] was (1) fallow, (2) tilth, (3) breach. After summer fallow stirrings, each shift in turn was first brought to a fine tilth for a tilth crop, and then, when its stubble had been ploughed in, the seed for a breach (etch) crop was harrowed into the rough clods. The tilth field was for food and drink crops, mainly wheat, rye, maslin, bigg and fallow spring barley. The breach field was for feed crops like peas, beans, oats and barley, which, being sown on the stubble clods, was called 'brush' barley. Then the field was bare-fallowed and brought again to a fine tilth by twice or thrice-gathering the ridges. This laid the corn safe from winter rains, and the wheat or rye, or, in spring, the barley, was usually sown under furrow and ploughed in. If the tilth field were too lean for wheat, rye or maslin was grown, and in lighter soils peas were preferred to beans. The cultivators chose what crops they pleased in the tilth and breach fields; they could, if they were so minded, sow both to barley, or grow, say, hemp in the breach field. That part of the tilth field sown to wheat in one year would usually, when the course came round, be under barley, while oats, peas, barley and beans might similarly be alternated in the breach field. The acreages under these different crops were not necessarily either equal or constant, but the general upshot was in the nature of a sexennial crop rotation of, say, (1) wheat, (2) barley, (3) fallow, (4) barley,

(5) peas, (6) fallow. Moreover, in many fields a hitch crop of pulse, usually peas, vetches or lentils, was taken on that part of the summer fallow known as the hitching or the hookland, for the pulse was meant either for the horses to be hitched or tethered on or for cutting with a hook to make hay for them. Also, small parts of the tilth field were widely sown to garden peas for gathering, partly by the poor, on condition they refrained from picking elsewhere.

The intermixture resulting from promiscuous cropping was sometimes reduced by an order that all should sow their winter corn as near to each other as possible in the tilth field and, to this end, exchange their lands indifferently one with another.[272] It was, nevertheless, a highly variegated picture that the common fields presented. Whoever walked Wheatley Lower Field in July 1635 would have met with successive parcels of maslin, rye, wheat, grass, fallow barley and oats; and in the Over Field he would have seen grass, brush barley, oats, mixed pulse, beans, peas, vetches and hemp. Hemp was growing on one ridge between pulses and a grassy ley land and elsewhere between two parcels of barley. To some extent the crops were grouped together, but were often scattered at random, not only amongst parcels and ridges, but also between different parts of the selfsame ridge. Part of one ridge was in peas and the rest of it in barley.[273] As at Cold Higham, a single ridge might have one end in wheat, the other in rye, and the middle in maslin.[274] Diversity of crops within an unchanging field-course is illustrated also by Clayworth common fields towards the close of the seventeenth century. The first field had a succession of (1) wheat or barley, (2) peas, (3) fallow, (4) wheat or barley, (5) peas, (6) fallow, (7) wheat, barley or rye, (8) peas, (9) fallow, (10) wheat, barley or rye. The concurrent succession in the second field was (1) barley, beans or peas, (2) fallow, (3) rape fallow followed by barley or wheat, (4) barley or peas, (5) bare fallow, (6) 'white corn', i.e. straw crops as before, (7) peas, (8) bare fallow, (9) 'white corn', (10) peas or barley. Meanwhile the succession in the third field was (1) fallow, (2) wheat or barley, (3) peas or barley, (4) fallow, (5) 'white corn', (6) peas, (7) fallow, (8) 'white corn', (9) peas, (10) fallow. In 1688 one cultivator had, in the tilth field, five and five-eighths acres of wheat, three of barley, and five of rape fallow in preparation for barley; and, in the breach field, eleven acres of peas and three and a half of barley.[275]

In the Vale of Evesham, the Fen Country and parts of the

Midland Plain there was an extended plain-country course of (1) fallow, (2) tilth, (3) breach, (4) breach. In this course, which became the predominant one in the Vale of Evesham, there were two main alternative crop rotations, viz. (1) fallow, (2) barley, (3) pulse, (4) wheat, and, less usual, (1) fallow, (2) wheat, (3) barley, (4) pulse. Wheat was the chief winter corn, and the pulse crops consisted of beans, peas, 'frouse' (a mixture of many horse beans and a few grey peas), and tares sown with oats to climb by.[276] Less ploughing was needed for three crops in this course than in the plain-country three-field one. In the four-field course,

> the number of plowings, in four years round, is six: three in the fallow year; one for barley; one for beans; and generally one for wheat. The fallow is broken up after barley seed time; slitting the ridges down, by a deep ploughing. In the first stirring they are gathered up. On this second plowing, the manure is spread; and plowed under with a shallow furrow; which is likewise turned upward, to lay the ridges during winter. In the spring, they are split down for barley; and the next autumn gathered up for beans; and the ensuing autumn, again plowed up for wheat. Six plowings, in four years, for three crops and a fallow; four of them being upward, two downward of the ridges.[277]

In the Midland Plain, the best clay loams, as in the vale of Belvoir, were often in a four-field course, with barley or oats in the first breach field and peas and beans in the second.[278] Thanks to abundant fen fodder and pasture, it often happened in the Fen Country that heavy stocking and liberal mucking permitted highly productive field-courses. A four-field one was widely adopted, with wheat, barley, beans and peas as the main crops. As in the Vale of Evesham, there were two alternative rotations, pulse either preceding or succeeding barley. Some common fields were cultivated even more intensively in a five-field course of (1) fallow, (2) tilth, (3, 4 and 5) breach, the first and third breach fields being usually in barley and the second in peas or beans.[279]

At the other end of the scale, some plain-country soils, even in the Cheese and Fen countries,[280] were fallowed in alternate years and were thus in a two-field course, viz. (1) crop, (2) fallow, mostly bare, but part sown with a catch crop of pulse and oats. This course was often followed in the Midland Plain, especially in earlier times. In the Fen Country a two-field course was often preferred on account of the greater opportunities afforded for growing fallow and smothering crops.[281]

All these plain-country courses were admirably adapted to deep, cold, heavy soils that had to be gathered into high ridges for winter corn, especially when they could be folded by sheep only in the summer months. But the two-field course suited many different places in both the plain countries and the downlands.

The foremost practitioners of the two-field course were the common-field farmers of the Cotswold Country, where the original course was almost invariably (1) fallow, (2) corn, for which the fields were divided or grouped into two seasons,[282] with the proviso, however, that some soils could not bear even this mild yoke and went barren from time to time.[283] Thus at Sevenhampton, with the exception of Blackthorns Common, which lies between them, 'all other the feilds and commonable places ... are devided into two feildes, the one feild severall and the other feild common yearely and ... one of the said feildes conteyneth the Quarr Feld'.[284] In the trial of a customary tenant in Alderminster manor court for sowing hill land to wheat in two successive years, 'it is proved by evedence to the jury that it was never knowen to be but every other years land'.[285] While one half was being fallowed, the other was sown to wheat, barley or oats. All corn was thus preceded by at least one year's fallow, and the rotation might be: (1) fallow, (2) wheat, (3) fallow, (4) barley.[286] The choice of crops, however, was strictly for the individual cultivator. He might grow hemp if he wanted. The men of Idbury, we are told, were accustomed, 'if in wet ground the ways be made broad and the corne lost, to supply what is lost by soweing of hemp when the weather comes to be drye, and makeing the path narrow'.[287] In many townships with cornbrash soils, barley was grown almost to the exclusion of the less reliable wheat, rye and maslin, and barley bread was consequently for long the staff of life.[288] Part of the fallow field was generally sown to a hitch crop of lentils, tares, peas, or, on the cornbrash, of beans. All could equally well be tethered off or mown for provender. Hitching fields or 'in-quarters' were usually in the lower lands, nearer the farmsteads, but were shifted biennially. If necessary, the hitching field was temporarily redistributed to the various commoners, according to a rate determined by the size of their farms. In the seventeenth century there was a marked tendency for the hitching fields to grow in size, often to such an extent that it became convenient to divide them up into separate pea, vetch and lentil hitchings. In none of these did the land escape a bare fallow, for the crop was always off in time for cleaning cultivations in the late summer or early autumn; and generally the hitching fields were destined for barley or oats and could receive winter fallowings.[289] Thus all the land was bare-fallowed every year, for eighteen, twelve or six months.

Two-field courses, with ample hitchings, were followed also in the thinner soils of the Chalk Country. Here, again, the farmer grew what crops he liked in the corn field, varying their proportions as he saw fit, and often having more barley than wheat.[290] In the lighter of the soils below the hill in the Northwold Country, the prevailing course was (1) fallow, (2) corn, with the fallow all bare; but in the deeper of them, half the fallow was sown with vetches or other pulses, with a succession after the sort of (1) bare fallow, (2) barley, (3) fallow crops followed by cleaning cultivations in late summer, (4) winter corn.[291] Other towns with two-field courses were found in the Southdown, Oxford Heights and Poor Soils countries and the Norfolk Heathlands.[292]

The usual course in the deeper and better of the warm loams, however, especially the calcareous ones, was the three-field one of (1) fallow, (2) winter corn, (3) spring corn. This prevailed in the Chalk, Southdown and Oxford Heights countries, and in the Vales of Hereford, and was also found in parts of the Northwold, Cotswold, Poor Soils, Breckland, Fen and other countries, and, apparently, within and about the Pale of Dublin.[293] In the Chalk and Southdown countries, the winter corn was mostly wheat, partly rye; the spring corn more barley than oats; and the fallow partly hitched. Similar successions were employed in other countries,[294] but in the Vales of Hereford, beans were the favoured spring crop in the 'wheatlands', and peas and oats in the 'ryelands'.[295]

Just as the plain-country three-field course was stretched into a four-field one, so too was the downland one, by a repetition of the spring field, giving a course of (1) winter corn, (2) spring corn, (3) spring corn, (4) fallow. It was then usual to have barley in the first spring field and oats or some other summer corn in the second, which was often called the summer field. This four-field course was found in richer soils in the Chalk Country.[296] It was common in the Breckland, though with some special adaptations. Here still fallowing ('summerley') was everywhere preferred, lest the soil blow or scorch. Winter corn was mostly rye, and summer chiefly lentils, tares and peas. Thus at Foulden in 1620 the field-course was (1) summerley, (2) winter corn, (3) barley, (4) promiscuous crops of summer corn. At Weeting in 1604 the common fields had long been arranged in five 'shiftes, which they call feildes'. The first shift consisted of East Field and Shearham Furlong; the second of North Field, with the exception of a piece in its centre called Coxindeale Field or Furlong; the third, of West Field; the fourth, of South Field; and the fifth, of Coxindeale itself. The first of these shift fields was under the following course: (1) rye, (2) barley, (3) oats, barley or 'summer grain' at pleasure, (4) unsown to gather heart and to be folded with sheep. In the third year of this succession, known as the 'oat shifte', this shift field was sown no further than to a certain half-acre lying near Walsingham Way, beyond which the land was summerley two years running. The second, third and fourth shift fields were sown in the same succession as the first, the oat shift being similarly restricted. The fifth shift field, to wit Coxindeale, was never sown to rye but only to barley, and the succession here was (1) barley, (2) barley, (3) part barley, part still fallow, (4) still fallow. No wheat was grown because the land was too light and sandy. The still fallows were fed and folded by the fold-course flock, which, as the saying is, turned the sand into gold.[297] Finally, in the Norfolk Heathlands, at least, these downland courses had been stretched even further, giving successions like these: (1) summerley, broken up, stirred and sown on four or five earths (ploughings) for (2) wheat, (3) barley, (4) peas or vetches, (5) barley; (1) summerley, (2) wheat or rye, (3) barley, (4) peas, (5) oats; (1) summerley, (2) barley or summer corn, (3) summerley, (4) rye or wheat, (5) summer corn; (1) summerley, broken up and sown on four earths to (2) rye, (3) barley, (4) peas, (5) barley, (6) summerley.[298]

The Lancashire Plain townships usually had small common fields, in which most farms had their share. But cropping was not regulated in manor courts, whose ordinances were largely confined to such matters as scouring ditches, impounding strays, prosecuting encroachers and managing common pastures. Generally, when the field was being cropped, all sheep were in the charge of common shepherds, and from barley-seed to barley-mow even the tethering of beasts was prohibited. Besides the common herdsmen and bylaw men, the townships appointed pinders, moss-reeves and house-viewers. That common-field administration was present in all its essentials and almost all its manifestations makes it all the more remarkable that it hardly more than regulated common of pasture when the field was 'abroad' in the winter months and debarred it in the summer ones.[299] Moreover, as we have already had occasion to notice, the full and comprehensive Leyland town-field agreement of 1612 mentioned neither field-courses nor crop rotations.[300]

Yet this taciturnity is readily explained, for it was precisely the alternation of being 'abroad' in winter and cropped in summer that constituted the field-course. Between 1542 and 1547 John Moore kept account of the crops on his land, most of which lay in Liverpool town fields and the rest in closes in or near them. In both closes and common-field parcels, about half the land was sown to barley and a quarter to oats. In the town-field parcels only, vetches grew on an eighth of the crop acreage, and on the remaining eighth largely peas, nearly all grey, and only a few beans. In the closes or Lammas grounds no vetches were grown and nearly an eighth of the land was sown to rye or to a maslin of rye and wheat, and the rest mostly to peas and beans, separate or mixed, and sometimes with oats for them to climb by. Small amounts of wheat and buckwheat were also cultivated, and minute ones of bigg and linseed (flax). The point is, winter corn (wheat, rye, maslin or bigg) was grown in the closes, but never in the common fields, which were then 'abroad' to common of pasture. Winter corn in the town fields was not perhaps explicitly forbidden, but no one grew it, for it would immediately have been eaten off green by the common flocks and herds. In no part of the town fields was there any exact parallel to the field-courses found in other countries. Twelve-month bare fallows were never seen, but winter fallows were made every year. Whereas in most common-field courses all or most cleaning operations were concentrated in fallow years set aside for the purpose, in this 'one-field' course, as it is sometimes called, they were all done in the winter months. Every autumn and winter, all the land was fallowed, and every spring all the land was sown to crops. Fallow stirrings were carried out, but not usually on all the land; some was just left in stubble over the winter. Although the field was cropped relentlessly every year, spring crops exhausted it little. Barley was the cereal most grown, in distinct fallow and brush crops. Fallow barley was sown in a fine tilth after stirring and mucking, brush barley in land prepared only by the ploughing in of the stubble. The most usual succession was an alternation of fallow barley with some other spring crop, e.g. (1) fallow barley, (2) oats 'on brote', i.e. on the stubble brush, (3) fallow barley,

(4) brush oats or brush barley, (5) fallow barley, (6) spring vetches, and so on. The field-course may thus be defined as (1) winter fallow mucked and stirred for barley on a fine tilth for bread and drink corn, (2) still winter fallow depastured for a spring crop, partly of feed barley, sown on the brush. Unlike the closes, the town fields had their arable all in permanent tillage, and unlike the closes they were seldom or never marled.[301] What permanent common fields there were in the Welshry of Wales had a field-course similar to the above.[302]

Finally, any of the multifield courses described above could be converted into 'every year's land' either by extending the hitching to the whole of the fallow field, or by growing smothering crops in all of it, leaving no long bare fallow. In East Norfolk and the Northdown Country, already by the high Middle Ages, it had become practice to grow hitch crops ('podware') in almost all the fallow fields.[303] 'Every year's land' of this kind was also found in the Chalk Country, at Amesbury, Stoke Farthing and elsewhere. 'The common field at Stoke Verdon', says Aubrey, 'is sowen every yeare according to husbandry: never lies fallow: it is very good land, lies desk-like to the South.'[304] In the Vale of Evesham such every year's land was cultivated without any formal agreement between the occupiers, with no sheepfold, no fertilisation except by leguminous crops and farmyard manure, and with barley the nearest thing to a cleaning crop.[305] Every year's ground was frequently met with also in parts of the Vale of London where the soils were free-working and muck from the city streets, stables and cowsheds was readily available.[306] In the Midland Plain, Houghton Regis (which had a staging camp for drovers and livestock and also maintained teams of horses to help carts and wagons negotiate Chalk Hill), Higham Ferrers, which was a populous town, and Catesby, Great Houghton and Wollaston, all had much every year's land.[307] Peterborough Bon Field had an endless succession of corn and pulse,[308] and elsewhere in the Fen Country corn often alternated in the heavier soils with flax, in the warmer ones with hemp, whole-year bare fallows being altogether avoided. Flax and hemp were smothering crops and prepared the land for grain. Thus in Epworth common fields the practice was to 'sowe every yeare, one yeare hempe, the nexte yeare barley, the nexte hempe againe, and the nexte rye'. The fallow was ploughed in winter and sown in spring. By July or August the flax and summer-ripe hemp would be off the field, which was then accorded two or three fallow stirrings and manured in preparation for wheat, rye or barley.[309] Every year's ground could exist only in favoured soils and situations with abundant supplies of manure. In such grounds the succession of crops was endless. Pulse and corn, or market-garden and farm crops might be alternated, but, in the absence of long bare fallows, the land became foul unless weeds were checked by hoeing, by weeding, or by smothering crops of mustard, hemp or flax.[310]

There was thus no uniform course for common fields. In all the different countries one field-course or another predominated, but in no one, unless the Lancashire Plain, did all common fields follow the same course. Often all the fields in a particular town were under one and the same course, but just as frequently one part had one course and another another.[311] On the bounds of the Midland Plain, for example, Barton-in-the-Clay and Wendover both had numerous common fields, some in the vale and others on the hills, under entirely different courses in the two different situations.[312] Potton Regis had a special Heath Field with its own separate pinder.[313] Similarly distinct rye or heath fields existed at Greens Norton, West Haddon, Hampton-in-Arden and other places.[314] Rearsby had a separate Sand Field,[315] Stratton Audley two groups of stone and clay fields.[316] A township might, like Higham Ferrers, have one of four topographical fields as every year's land and the other three in an ordinary three-field course.[317] In Barford, some common field was in a two, some in a three, and some in a four-field course.[318] Several towns in the early modern Vale of Evesham had both two and three, and some also four-field courses, and some in the Fen Country both three and four-field ones.[319] Similar diversity was found in the Chalk Country.[320] At Easton and Wroughton, for instance, the hill ground was sown every other year, while the lower fields were in a three-field course.[321] The hill land near Hungerford also had its separate two-field course.[322] At Henwick, in the middle of the seventeenth century, some of the common fields were fallowed every third year, but not all.[323] In Mere, some of the common field was in a four-field course and some in a two-field one.[324] In Amesbury, three of the fields were in a four-field course, one in a two-field, and three were 'every year's ground'.[325]

Nor was common-field husbandry unchanging and unbending. One field-course might be superseded by another. In the Vale of Evesham, as the soils were ameliorated, a township might rearrange its seasons and pass from the originally widespread two-field course, or from a three-field one, to the four-field course characteristic of the Vale in early modern times. In the early seventeenth century, for example, the Woodmancote common-field farmers concluded that three-field courses were 'unhusbandly' and decided to change to a four-field one, conforming to the 'usage and custome of the Vale of Evesham, within the precinct of which country they lye.'[326] In the Midland Plain, in 1615, Marston Priors made a similar change from a two- to a three-field course, deciding 'that the feildes belonginge to the said towne shoulde be devided into three partes for twelve yeares, one for white corne, one pease and one fallowe'. Kelby fields were recast into three in 1693. Yet another Midland town, Grandborough, had resolved, early in the seventeenth century, that 'there shalbe three feilds for three yeres and then as it shalbe thought meete'.[327] Many other towns in the plain, or on its borders, changed over from two to three fields before 1550, not a few before 1350, suggesting that the change from two- to three-field courses was proceeding gradually from Anglo-Saxon times onwards. This improvement was important, but hardly revolutionary, especially if the old two-field course had included a large hitching field.[328]

The Midland Plain apart, the evidence for changes from two- to three-field courses is slight and somewhat confused, being confined chiefly to border towns that partook of the practices of two countries. The only credible instances cited by Gray refer to South Stoke and Piddleton in the thirteenth century.[329] The former town was on the borders of the Chiltern Country, as it later was, and the latter on those of the Chalk Country. This suggests that the changes were parts of a process by which distinct farming countries emerged.

By far the greatest changes in field-courses were those in the Cotswold Country. From about 1560, in a general movement, town after town recast two fields into four quarters, or arranged some similar grouping, in order to introduce a four-field course. In this 'three-crop land', one quarter was fallowed and the others sown successively to wheat, barley, and pulse.[330] But only in the deeper and more ameliorated soils could this change be made. Sometimes, when the lower lands had been thrown into four quarters, the hill fields continued in a two-field course, making six quarters altogether.[331] In some towns, like Aynho, while some of the field was set aside as three-crop land, another part was made two-crop land, in a three-field course.[332] Other towns adopted a single three-field course, and yet others not only retained the two-field course, but even eschewed a hitching field, on the ground that it detracted from the succeeding crops.[333] Nevertheless, the hitchings were generally continued, and even extended, in what was sometimes called the 'in-quarter'.[334] Where the lower lands were thrown into quarters, while the higher ones continued in the old course, the latter were sometimes designated 'outlands' and the former 'home fields'.[335] After such rearrangements, the coincidence of topographical and shift fields was even rarer than before. Field-courses were based on quarters, not necessarily signifying fourth parts, and required readjustment from time to time.[336]

We have seen that field-courses were designed and chosen to suit particular places at particular times and not for any reason arising from tradition or inertia. We have also seen that within a given field-course, the individual might choose what crop he grew. What is perhaps not always realised with perfect clarity is how wide this choice was. Common-field cultivators were completely free to grow flax, hemp,[337] or saffron, for example, as long as these could be fitted into the field-course. Saffron stood several years in the field, and just how flexible common-field cropping arrangements could be is perhaps best illustrated by a description of saffron cultivation as practised about Saffron Walden in the Chiltern Country, and, to a lesser extent, in the Norfolk Heathlands, particularly round about the Walsinghams.[338] The bulbs were planted in small temporary enclosures a few roods or acres in size and defended against animals, especially sheep and hares, which would have eaten off the leaves in winter. Saffron was set about midsummer, on a tilth of three ploughings composted with farmyard manure and pigeon dung. During the second and succeeding winters, the plant remained in leaf, and in the second and succeeding summers, raised its head. After this it was taken up, except in the richest grounds not earmarked for anything else, where it might be left for a fourth year. In the second, third and fourth summers, the first blades died away altogether, but in September the flower opened its six bluish purple petals. In the midst of these petals were the chives from which were obtained condiments, perfumes and medicines. The chives had to be picked at sunrise on the first day of their appearance, to forestall the birds, the bees and the withering sun. Afterwards the flowers were plucked and thrown on the dunghills. All the first season, any cultivation other than ploughing could be done without harming the saffron. In the second and third seasons, saffron disrupted husbandry no more than successive tilth and breach years would have done, except that sheep had to be excluded in the second winter when the land might otherwise have been folded for the breach crop. To compensate any who preferred not to cultivate saffron, those who grew it had only to forgo their own rights of common for sheep in proportion to the small quantity of saffron ground enclosed from the field.[339] When we see a difficult crop like saffron accommodated in common fields, it is no longer a cause of wonder that hemp was so easily grown in them.[340]

Two opposing schools of thought have dominated discussions of common-field cultivation. One consistently and insistently ascribed the downland three-field course to all countries whatsoever, or at least to the plain countries and especially the Midland Plain, where it was never followed.[341] The other school arrived at the perverse conclusion that there were no regular courses in common fields, but only an infinite flexibility that endowed them with most of the advantages of severalty, for, as was only to be expected, a comparison of the downland three-field course with the crops actually grown in the plain-country three-field course revealed a lack of coincidence between the two.[342] The root of the errors of both schools of thought thus lies in an aversion from the particular and a tendency to over-generalisation. The whole issue has been further bedevilled by the confusion of field-names with field-courses. Once the necessary distinction has been made between the two and it is understood that field-courses varied from country to country, from town to town, and from field to field, the 'problem' turns out to be a mare's nest. The degree of tolerance in the choice of crops and rotations was not the same in all three-field courses. In the downland three-field course, 'horsemeat' might be grown in the hitching field and the cultivator might further choose between wheat, rye and maslin for the winter-corn field and between barley, peas, oats and dredge in the spring field. In the plain-country three-field course, the cultivator might grow any crop he pleased in the tilth field and in the breach field; he had only to stick to the tilth and the breach. In a two-field course, the cultivator was bound to fallow half his land every year, but in the other half he could grow whatever he wanted. In the two-field course he and his fellow townsmen followed, Robert Loder was free to make a rational commercial choice between wheat and barley. He would have had the same freedom under the plain-country three-field course, but not so if the downland three-field course had been operated in Harwell. But then again, if it had been, he would not have wanted a choice, for he would have had both crops. Four-field courses further restricted a farmer's choice of crops, but likewise lessened the need for him to choose. In the Lancashire Plain course, anything could be grown in spring, but nothing in winter. In every year's land, everything depended on what field-course it had been derived from by the suppression of the year's bare fallow. There was, in short, considerable flexibility in the choice of common-field crop rotations, but all inside the rigid framework of fixed field-courses, which were regulated and enforced with the utmost exactitude and rigour, with all the sanctions at the command of manor courts, and in defect of these courts, by any and every possible legal means. Maximum regulation of field-courses, maximum freedom in cropping, was ever the rule in common fields.[343]

Despite variations in cropping and tilling, common-field courses became in no way irregular. In the Chalk Country, crops and cultivations might vary, but field-course routines were enforced. Even the size of the hitching was usually restricted, lest horsemeat oust sheep from the fallows. According to articles drawn up by the Collingbourne Ducis tenantry in 1593, Jeffereys, the demesne farmer, was forbidden by custom to plough more than thirty ridges in the spring or summer field, and those only for spring-sown horsemeat. All the rest was to lie fallow until the wheat season and meanwhile to be folded by the freeholders' and copyholders' ewe flocks and the farmer's hog flock. The field-course and its seasons had to be maintained and cross-cropping prevented. In 1593 Mill Ball Field was wheat, East Field barley, and Middle Field 'somer fyeld and gratton' for the sheep of the 'worklanders', 'monday-landers' and freeholders. From barley-mow 1593 to Michaelmas 1594 this fifty-acre field was to have 342 sheep folded on it. When Jeffereys prepared to sow barley and other Lent corn in part of it, the township was figuratively up in arms.[344] By a consensus or great majority, the tenants could, had they wished, have changed the custom, but their main concern was to preserve the field-course and the routine of sheep-and-corn husbandry. Cross-cropping was opposed because not only might it lead to land being ploughed out of heart, but would also disrupt folding arrangements. Whatever field-course had been agreed, had to be enforced. Thus in 1605 the

Enford jury *presentant quod Thomas Downe citra ultimam curiam arravit unam acram terre jacentem in quodam campo ... contra antiquum cursum* – 'present that Thomas Downe since the last court ploughed one acre of land lying in a certain field ... contrary to the ancient course'. He was amerced and forbidden to repeat the offence under pain of £1.[345] In the manor court of Mere in 1598 Christopher Aubrey, the demesne farmer, was presented for sowing wheat in the barley field, when it should have been sown only in 'barley seede tyme'.[346] Thus although crop rotations were not regulated, they had to conform to field-courses. In the Midland Plain, no less, common-field courses continued to be rigidly and meticulously regulated right up until the eighteenth century or later. Just the same is true of the Cotswold and all common-field countries.[347]

After 1650, as new crops were introduced into more and more of them, the courses in common fields underwent more frequent changes. Clover and 'seeds' entered the common-field practices of the Midland Plain, the Vale of Evesham, and the Fen, Chalk, Southdown, Oxford Heights, Cotswold and Northwold countries; sainfoin those of the Cotswold and Chiltern countries; turnips those of parts of the Midland Plain, the Vale of Evesham, and the Fen, Chalk, Cotswold and Oxford Heights countries; and potatoes those of the Fen Country. Carrots were cultivated in the common fields of one Vale of Evesham township as early as 1656 and clover in those of a Midland one no later than 1662.[348] In the next hundred years more and more field-courses were changed to accommodate the new crops. In the Cotswold Country, the two-field course, which had already given way to a four-field one on 'three-crop' land, was now superseded by a system of four or more 'quarters'. The 'partitions' of the Oxford Heights Country performed the same function. In the Fen Country and the Vale of Evesham, the four-field courses were easily adapted to take the new crops. In the Midland Plain, some townships adopted improved four-field courses, some six-field courses, and some nine-field ones, while still others contented themselves with growing clover in the fallow season of the old three-field course, so often converting it to every year's land. In the Chalk Country, the three-field course was sometimes retained in a modified form, and sometimes replaced by a four-field one, or an old four-field course was reformed. In consequence of all these and similar changes, crop rotations became much more complex. The new field-courses accommodated a wider range of crops, but the courses themselves were no less inflexible than the ones they replaced. It might have been supposed that these new crops would have put an end to the rigid enforcement of field-courses; but not at all. On the contrary, their introduction necessitated new and often highly intricate courses based on shift fields whose bounds were inclined to be fluid; and this rendered essential ever more complicated by-laws and customs and the even stricter regulation and enforcement of field-courses.[349] In the common fields newly laid out by some enclosure commissions, legally binding field-courses were set out and laid down in the awards themselves;[350] but some remnants of common fields left after enclosure awards were suffered to go adrift without any adequate system of field-course enforcement, and neighbourly agreements could not always suffice for the purpose. An important cause of the decline of the enforcement of common-field regulations, and of the consequent fall in standards of husbandry that often occurred, was the decay of lordship and of manorial institutions.[351] It was to remedy the ill effects of this decline that an Act of Parliament in 1773 allowed field-reeves elected by common-field farmers to enforce by-laws passed by town meetings.[352]

CHAPTER 4

Field and fold

Common fields and common folds were almost inseparable,[353] for in order to fertilise their arable the little common-field cultivators had to put their sheep into common flocks. To this general rule there were some exceptions. Those parts of the field nearest to the farmstead, and the whole of small fields, like many in the Lancashire Plain, and of the infields in Wales and elsewhere, could be fertilised by spreading farmyard manure conveyed to them by dungpot or dungcart. Temporary common fields in the mountainous countries could be brought into cultivation either by tathing with all the available livestock, or by paring and burning, or by rib-ploughing. Nevertheless, for the bulk of English common fields, common sheepfolds were essential.

In the Chalk Country, the man with only a few score sheep could not fold his arable with these alone. As most of the farms had pasture for fewer than a hundred sheep, common flocks were inevitable.[354] To employ such a common flock, the tenants in a township had to treat the whole of all their lands as one single unit for all that concerned the flock and the fold. Hence they had to consult together, abide by mutually agreed regulations, and form a common purse.[355] That a small flock was worthless for folding and that folding was the chief purpose in forming large flocks, is further demonstrated by the fact that even some substantial farmers, with sizeable flocks of their own, still preferred to entrust some or all of their sheep to the common shepherd. At Amesbury Earls in 1587 the demesne farmer's hog flock, the town flock, and that of one of the freeholders were all merged into one for the purposes of folding.[356] In 1633 the tenants of Alvediston agreed to confine themselves to one or two shepherds chosen by a majority voice.[357] What a farmer with a few hundred sheep lost by putting them into a common flock was more than compensated for by the improvement in his crops. 'Such is the power of invention, when urged by natural necessity, that even the lowest class of farmers are enabled to keep sheep, and fold their arable lands, with a degree of propriety.'[358]

Common or tenantry flocks were managed by common shepherds under the tenantry's orders, which were promulgated in the manor courts and received the consents and signatures of all the homage. As exemplified by those in Heale court in 1629, such orders laid down the shepherd's terms of employment and duties and the tenantry's obligations, including the supply of hurdles and hay. Non-compliance by any tenant was penalised by forfeiture of all benefit of fold. Two men who failed to provide their quota of hay that year were deprived of the fold, and it was ordered 'that Robert Atkins in regard of his insolencye to the lord of this mannor nowe in open courte and for refuzinge to obey the orders of this courte, shall not from hencefurthe have the benefytt of the flocke to come uppon his lande'.[359] The records of other manors show similar regulations and agreements.[360] In 1622 every inhabitant of Kingsworthy was ordered to provide ten hurdles for each yardland throughout the year, and, on warning from the shepherd, to carry these hurdles 'from land to land'. When, in 1609, Thomas Poynter neglected to provide his hay for the flock and fold, he was amerced thirty shillings in the manor court of Brigmerston and Milston. Shrewton township ran two common flocks, of which one was reserved for hogs, crones and culls. When hay was not forthcoming, the shepherd was empowered to distrain on the defaulter's sheep. At Wylye court in 1633, West End sheep commons were stinted at forty to a yardland and a rate made payable to three men elected to 'imploy the said money in buying of hay for the flock'. A similar common stock for winter fodder was raised in Amesbury Earls by a rate of one shilling for every twenty sheep. Defaulters could expect no hay, and what this would mean on the bleak downs in the depth of winter may be imagined. It was their sheep perishing for lack of hay that induced the people of Shrewton to agree to common orders in 1598.[361]

Thus each common-field cultivator had his land dunged in turn acre by acre, and, since his parcels were dispersed and intermingled, received in his proper turn a fair share of the common fold he helped to supply.[362] Men with no sheep could nevertheless have their land dunged by the common fold on payment of a small sum. The custom in Amesbury Earls was for the fold to 'goe over all the townefeyld not skyppinge any mans land'.[363] In the three-field township of Winterbourne Earls, the 'tennants foulde goeth adead all over two of the tennants feilds once every year'.[364] In other words, the fold went over every acre in turn, and taking unfair advantage by 'turninge the fold out of his coors' was a serious offence.[365] The tenants of All Cannings had three acres folded for every forty sheep leas and at that rate, and everyone enjoyed the fold equally as long as it remained in the field.[366] The jury of Kingsworthy ordered 'that after the summer course of the fold be yearely ended, then every yardland shall have two nights of the wheate season course of foldinge uppon whole land and the resydue uppon corne. And when the fold by course cometh to anye mans land, then if he shall refuse to take his night, he shall quite loose it for that turne.'[367] It was important too, as stressed in Brigmerston and Milston court in 1610, that the fold should enter the fields at opposite ends alternately, in order to compensate for inequalities arising from the irregular dispersal of parcels.[368] Thus, only hindered in earlier times by fold-suit, i.e. the special faldage rights belonging to some manorial lords,[369] the tenantry arranged their folds to their common advantage and took all necessary steps to integrate them with their cultivations. It was ordered at Alvediston in 1633 'that whosoever shall fallow any lande in the sommerfield until it be within a weeke of middsommer yeerely (unless the same land be dunged with the pott or fold) or in the barley field untill after Christmas day yeerely (except it be dunged with the fold) shall forfeit for every acre soe fallowed in either fielde x*s*'.[370] Similarly, in 1726 and 1744 there were orders in Crawley against ploughing any tillage in the summer field until the fold had been over it.[371]

In the Southdown Country, too, common-field farmers folded their tillage by putting their sheep into common flocks and folds under common shepherds who moved the folds over the fields acre by acre.[372] The importance attached to this fold may be judged from the Wroxall by-law forbidding the commoning of any sheep not penned on the town tillage.[373] Common flocks and folds necessitated close co-operation, as may be seen in the regulations for the tenantry flock at Berwick in 1721. About a hundred acres, or upwards of seventy ridges, had to be folded each year. The sheep stint on the commons and 'laines' in winter was three for each bullock depastured on the leas, and in summer, four for every acre of 'greenland' or ley land in the fields, plus two for every acre to be fallowed. In addition, even those without land in the field were allowed to common three sheep each, provided these were folded along with the others. Three out of every four sheep put into the flock and fold had to be breeding ewes. As this was a township straddling the escarpment and the arable foreland, when the 'maam' (chalky clay loam) was miry, brushwood walkways were laid for the sheep where they passed to and from the downs. For this purpose each farmer had to supply faggots in proportion to the number of his sheep. The supply of hurdles for the fold and the number of nights' folding enjoyed by each man were similarly rated. Had a farmer not received all the nights due to him by the time fallowing was completed, he might take the remainder wherever he chose. If, despite the faggots, the way became too miry for the sheep, they were either to be left on the downs or put into the home closes. Otherwise, to take the fold off its appointed course was an offence entailing severe penalties.[374] Much the same system of common folds obtained in the Oxford Heights Country as long as common fields continued. In Bromham manor court in 1583 it is ordered that three sheep may go in the common fields for every acre of land there and 'yt ys agreed to have common fold provided amongst them to runne over all theyr land'.[375]

But, despite the prevalence of sheep-and-corn husbandry, common folds were rarely if ever organised in the Northdown Country, and in the Chiltern Country only in the larger common fields.[376] These countries, being so much enclosed, were not generally suited to this system of managing folds.

In many countries that had common fields and folds, sheep could not be folded in the winter months, and then the common fold could not be organised on a permanent basis as it was in the downlands. This was true of the Cotswold Country,[377] and even more of the plain countries, most noticeably of the common fields of the Midland Plain. Sheep were folded at night on the tillage behind hazelrod hurdles ('fold fleaks'), the fold being moved daily to fresh land. Common-field cultivators attached great importance to the fold and confined common rights to sheep intended for it. When sheep agistments were allowed, they were restricted to sheep folded on the tillage, and the same condition attached to the letting of common rights. It was chiefly for the sake of the fold that common flocks were formed. During the folding season, by-herds of sheep were forbidden and sheep could be removed from common flocks and folds only under the most exceptional circumstances.[378] One peculiar feature of the sheepfold here was that common flocks were not usually permanent, but specially constituted in spring for the duration of the folding season only, because folding was not pursued in winter and some farmers only took agistments or bought in sheep just for the season. Indeed, in the vale of Aylesbury, it was common practice to hire flocks from the neighbouring hills under the management of their itinerant flockmasters. Another peculiarity in the plain as a whole was that there was usually not one single common flock and fold, but a group of several flocks, perhaps as many as eight, some formed privately and some by voluntary association, and yet all under public control. A common flock might be set up by a farmer and a few cottagers, by a few neighbours in the field, or by a group of the petty subtenants of the same landholder, so each flock and fold would contain only between one and three hundred or so sheep. A hundred or a hundred and fifty could fold two or more acres a month or ten in the course of a season.[379] Folding flocks were small because they rarely enjoyed great sheepwalks and usually had only small pieces of common pasture, and as they fed in small numbers, it was easiest to fold them in the same formations. Although these small folds took several nights to fertilise a single acre, their combined action would have sufficed to fold a large field, if only the season not been so short.

In the light, dry soils that could be ploughed early in spring, folding might commence immediately afterwards and continue until as late as the middle of November; but in the clays and clay loams, fallows could not be broken up until about April or May. Only then did folding start, and it could not go on later than September or October. Moreover, the most valuable folding came only with the first stirring of the fallows, usually in June. In all the heavier soils, folding was impossible when the fields were wet, not merely because the sheep would have been in mortal danger from rot, but also because they would have poached the land and ruined the subsequent crop. Furthermore, they could not be folded on the tilth field in winter, for the land was raised in high ridges to protect the wheat crop and the barley land from the winter rains, and had sheep been top-folded on the former or folded on the latter, they would have sought shelter in the intervening water-furrows, where they would have rotted the sooner, or would have escaped through the furrows and under the hurdles. The folding season, therefore, was restricted at the most to seven or eight months, at the least to four, and on the heavier lands the full benefit of the fold was felt for about three months only. This meant a single flock could often fold no more than half a dozen acres a season, and that even all a township's sheep, numbering, say, one thousand, could hardly fertilise more than one or two hundred acres. Most of the fold's benefit would have been lost, had the land not soon after been ploughed and sown to corn; and since folding was confined to the summer, it could only prepare the land for winter corn. Wheat, rye and maslin were, anyway, the crops most highly regarded in the common fields, and it was in the natural course of things that the best manure – that of the fold – should be reserved for them. Moreover, without the sheepfold, good winter corn could hardly be grown, for the land needed not only the dung but also the tread of sheep, since fallow stirrings reduced even heavy clays to powder in summer. Every circumstance, then, conspired to induce the general practice of folding for winter corn and mucking the barley lands of the tilth field with farmyard manure. This meant the acreage under winter corn was always limited, but most severely on the heavier soils, which were most apt for wheat but less suited to the fold.[380]

In the Vale of Evesham the spreading of farmyard manure was often supplemented in summer by folding, and common-field farmers kept sizeable flocks. But, although a few townships had spacious sheepwalks, most had little or no pasture for their flocks. Rye was occasionally grown for them; otherwise they had little beyond what was found on the fallows, stubbles, leys and balks, so scarcely a blade of grass or even a thistle escaped them, and this made for clean crops.[381] The deeper a vale and its soils and the wetter its climate, so much the less practicable was sheepfolding. Sheep folded in the Fen Country and the Western Waterlands[382] and in the depths of the Butter Country were often in dire peril.[383] The fold was hardly seen in the Cheese Country, for the tillage was unsafe for sheep when cold and wet and they would have spent most of their time sheltering in, or escaping by, the deep furrows.[384] And the Vales of Hereford provided the extreme example of a country with common fields but with no sheepfolds, no common flocks and no common shepherds. Instead, the sheep were cotted at night and their muck and compost carted out to be spread on the fields.[385]

Common folds flourished best where a large number of common-field farmers either occupied both the tenantry land and the demesnes, as they frequently did in the Midland Plain, or else co-existed with capital farms, as they usually did in the Chalk and

Southdown countries.[386] But where the common-field farmers occupied a relatively small share of the land and were rather less numerous or less influential, there was a marked tendency, especially in the high Middle Ages, for common flocks and folds to be subordinated to those of the demesne farmers or lords of manors, who might exercise rights of faldage and compel the tenants who owed fold-suit to put their sheep with the demesne flock and fold, or, looked at from the other point of view, might make the services of their own shepherds available on advantageous terms to the petty farmers. What the exact terms of these joint arrangements were depended on particular local cicumstances and varied between compulsory orders, payments for exemption, and free and easy co-operation with informal joint control.[387] At their liberal best, in the Northwold Country, such folding operations, under the shepherd employed by the lord or his demesne farmer, were described by Henry Best as follows:

The last ende of our lambs, being just 20 in number, and most of them sheeringe lambs, exceptinge some fewe other weake and younge lambs, went to field on Munday the second of May. From Lady-day, that our sheepe went first to field till Tuseday the 26th of Aprill and morrowe after St Marks day, the townsfolkes sheepe and ours wente togeather, and on nights weare carryed downe and layd att Hugill hill and the East dale bottome; but on Munday the 18th of Aprill the townsfolkes spoke to the shephearde that they should then hold of theire haver, which was sown in the middle field betwixt Killam gate and the Dale browe, for till that time they wente allmost as usually over the haver as the other lands. Whearefore, on Fryday the 22nd of Aprill, it being a wette morninge, wee sent our folde to field and sette it on the Spellowe flatte, and on the aforesayd 26th of Aprill beganne to folde. Wee sente 48 barres, wheareof 11 weare sette in eyther rake and 9 att eyther ende; then weare there as in every folde 8 corner barres. The number of the sheepe that weare folded weare 14 score and 17 olde sheepe and 6 score and two lambs ... Our shepheard took 3 lands and a halfe downe with him att a time, and soe made an ende of the flatte att twice goinge up and twice comminge downe.

As soon as the fallow had been fold-mucked in summer, it was stirred by up to four shallow ploughings for winter corn.[388]

In the Breckland, in the Norfolk Heathlands, and, in earlier times, here and there in East Norfolk, instead of mere faldage, which conferred the right to fold other men's sheep on one's own land, gratis, the great proprietors enjoyed fold-course rights, which allowed them also to fold their own sheep on other men's land, gratis. A fold-course consisted of four parts: the fold itself, the flock with faldage rights, a sheepwalk in severalty, and the right to feed the demesne flock, augmented as it as by other men's sheep, on all the common-field land when it was in shack, and when the field-course allowed no corn to be grown. The manorial tenants owed fold-suit (fold-soke) and, unless they bought licences of exemption, were bound to send their sheep, lambing ewes excepted, to the lord's fold, that he might have their 'tash' or tath, and to give precedence to the lord's flock in feeding shacks and summerleys.[389] Under this system both the lord's and the tenants' flocks were put together under a shepherd appointed by the lord of the fold-course or his farmer. The fold-course extended over the demesne heath, which was in severalty, the common heath, if any, the demesne arable, and, at certain times, over all the common-field land. There is 'fre fould-course', we are told, on the common field 'when yt lyeth somerley in the sommer tyme and in tyme of shack over all the filds savynge the same fild which is sowne with wyntercorne.' The flock-master had the right to fold the whole flock on his own land and to mow as much grass from the common-field balks and ollands as he needed to keep the sheep through the winter, but he had to shoulder the cost of shepherding.[390]

Since a single town often had two or more manors, and splitting a common-field town between two or more lords necessarily entailed a division of the tenants, together with their entire tenements and their scattered lands, the resulting boundaries between lordships were inevitably ragged. Nevertheless, each manor had its own fold-course, delimited, usually, by natural boundaries, which often coincided with those of the precincts, or by mounds marked by mere-stones.[391]

The fold-course in no way excluded the tenants' sheep from the field; it only excluded other flocks from the sheep-course. Within the fold-course limits, in addition to the liberty enjoyed by the fold-course's owner, the freeholders had 'cullet rights' or 'trips', for, say, from 60 to 200 sheep between them, or at the rate of one sheep to an acre of shack, these 'cullet', *collecta* or common flocks to go with the lord's and be kept by him at his own charges, except that the tenants had to allow hay for the winter. In return, the flock-master had a right to the 'tash' of the sheep at all the accustomed times.[392]

It so happened that the lord of the manor and owner of the fold-course or courses was often also the only substantial occupier in the township, or else that the fold-course and the demesne farm were both leased to a farmer who was also the chief freeholder and copyholder, so that the right to 'trips' sometimes had no practical result.[393] Some townships, however, had independent 'cullet' flocks, under separate common shepherds, that either went over the summerleys and shacks when the great fold-course flocks had finished with them, or had special sheep-courses of their own.[394] Tenants could often also have their land 'tashed' for them in return for small payments to the sheep-masters, whose shepherds usually covenanted not to tash other men's ground without their employers' express consent or, at least, not until all their own ground had been folded.[395]

It would be wrong to regard the owners and lessees of fold-courses merely as sheep-ranchers with interests diametrically opposed to thoe of farmers. The capital farms that accounted for most of the agricultural output were conducted on the principle of sheep-and-corn husbandry.[396] For instance, George Nonne of North Pickenham had, in the spring of 1633, 480 sheep worth £200 and four dozen hurdles for their fold, but his seventy-five acres of winter corn alone were already worth £100. Most of his land would have been sown to spring corn and he maintained nine plough-horses where no plough needed a team of more than three.[397] In 1660, Luke Constable esquire of Swaffham Market had 900 wethers in the heath and 460 ewes, worth altogether £630; and his shear of wool was valued at £197; but his standing rye was worth £40 and his granaries of corn £275.[398] Occupiers of demesne farms had economies similar to these; but great landowners had many fold-courses in hand and were interested not only in farming and farm rents but also in sheep-grazing pure and simple. Between 1545 and 1548 Roger Townshend had nearly four thousand sheep in four flocks at East Raynham, Kipton-cum-Helbroughton, Shereford and South Creake, each under its own shepherd. From the ewe flock at Kipton, hogs were drafted every year to South Creake, and at Shereford and East Raynham the flocks were of wethers. Each year Townshend sold about a thousand sheep, mainly wethers, and between two and four hundred stones of wool, which fetched between £36 and £114. From 1545 onwards, however, the Townshends kept the demesne arable of Barmer in hand and in 1620 were obviously extensive corn-growers as well as sheep-masters.[399] In 1561–2 Sir Richard Southwell had eighteen distinct flocks. Two of them were at Burnham Overy and Burnham Hoggast, and the others at Walsingham, Great Bircham, Bircham Tofts, Spixworth, Rudham, Weasenham, Cressingham, Ringstead, Barsham, Tottington, Threxton, Morton, Brancaster Staithe, where sheep were fatted on the saltmarsh, and at Brancaster Westground, where a fold-course was hired. This was a business with 17,000 sheep. In one year the surveyor of sheep accounted for an income of £1,378, of which just over £500 came from wool, almost £525 from lambs, about £237 from muttons, nearly £100 from store ewes, about £11 from fells, and £7 14*s* from payments for 'tashing' done for farmers. But the profit of the sheep-flocks amounted to only £755 out of a total of £3,208 accounted for by the receiver-general. Most of the estate's income came from farm leases, so even if Southwell kept no home farm in hand, still he had a great and abiding interest in agriculture.[400] Moreover, when landowners leased out their fold-courses, it was often together with a demesne farm, and sometimes also with a flock or the tash or tath of a flock, so, one way and another, pastoral and arable interests were combined.[401]

Nevertheless, great landowners tended to be great sheep-masters and their interests were somewhat opposed to those of the tenantry. The restrictions on cullet rights started in the fourteenth and fifteenth centuries, were multiplied and tightened in the years 1547–9. Betwen the sales in 1547 and those in 1548, the price of wool from 'Norfolk' sheep doubled. In 1545 Townshend sold 220 stones of wool for £36 13*s* 4*d*; in 1546, 282 stones for £47; in 1547, 342 for £57; and in 1548, 341 for £113 13*s* 4*d*. The increase was due largely to expansion of the flocks: by 1548 the number of sheep had increased by 480, about 12 per cent. Meanwhile, sheep profits were, in 1545, £99; in 1546, £111; in 1547, £126; in 1548, over £144.[402] So, whereas wool receipts had been less than half the sheep profits in 1545–7, in 1548 they made up three-quarters of them. For a brief period, then, wool became exceptionally profitable, and it was good business to sell fewer sheep and lambs, build up flocks, and sell more wool. This is exactly what Townshend, and probably others, did. Hooper thus reported truthfully to Cecil in 1551, about this part of the kingdom, 'All pastures and breeding of cattle is turned into sheeps meat, and they be not kept to be brought to market, but to bear wool.' Augmented flocks needed more walk and this could only be got by surcharging the common fields, chiefly by keeping the flocks in them longer, or by intruding into common pastures, or, even more effectively, by excluding cullet flocks. These were precisely the oppressions complained of.[403] Even after the crisis of 1548–9 and Kett's Rebellion, clashes occurred from time to time between sheep-masters and tenant farmers, who alleged encroachments on their commons and fields to the detriment of their tillage.[404] When Thomas Tusser was farming in West Dereham, he became involved in these conflicts and was later able to analyse them. He says in general, 'If shepherd would keepe them from stroieng of corne, the walke of his sheepe might the better be borne'. The worst oppression was in the tillage fields themselves, not in the common pastures and heaths, and the heaviest damage was inflicted on the winter corn. In Tusser's words,[405]

> The flocks of the Lords of the soile
> do yeerly the winter corne wrong:
> The same in a manner they spoile,
> with feeding so lowe and so long.
> And therefore that champion feeld
> doth seldome good winter corne yeeld.

In other words, the fold-course flocks were allowed to feed not only the ollands and the summerleys but also the corn, particularly the young winter corn. Now, the chief winter corn was rye, and rye makes excellent sheep feed in late winter and early spring. What happened, then, was that the tenantry sowed rye for bread corn and the flock-masters depastured their sheep on it. Had the sheep been top-folded sparingly on rank rye, little reasonable objection could have been raised; but they evidently fed the rye right off. In an extreme case, then, dire oppression might consist merely in allowing the sheep to feed too hard and low. Nevertheless, for the sheep to be in the rye at all, custom must have been broken, either by putting the sheep where they had no right to be or by sowing corn where the sheep should have been.

Just the same kind of conflict arose when tenants set about growing saffron in field land temporarily enclosed to exclude the flocks and stop them cropping the leaves and treading down the heads. Enclosure then worked in favour of the cultivators and against the sheep-masters. Hence the demand by the rebels under Kett in 1549 that enclosed saffron grounds be exempted from anti-enclosure legislation.[406]

After the defeat of this rebellion and a subsequent period of quiescence, the cultivators mounted a new offensive on the fold-courses. Some men allegedly trespassed on the fold-course sheep-walk. At Horsford in the early seventeenth century, Christopher Reeve, gentleman, and one Robert Bullayne (Bullen) were continually presented in the manor court, on informations preferred by the lord himself, for staff-holding on West and Black heaths and on the shack, i.e. for shepherding their flocks there. How this conflict ended is not known; the two parties are last heard hurling threats at each other in open court.[407] The more usual way of joining issue, however, was simply by sowing out of course. Changing field-courses and cultivating fallow and biennial crops deprived the fold-course flock of its accustomed summerley and shack and forced the flock-master to leave himself open to actions of trespass.[408] Saffron cultivation continued in the Norfolk Heathlands,[409] but the next open conflict of this type occurred at Great Breckles in 1586 and arose over the growing of corn. The flock-master alleged the defendants had sown rye in certain ridges in the field that should have been summerley, merely in order to deprive his sheep of their feed and to provoke actions of trespass at the common law. Despite the defendants' answer that they might sow and fallow as they pleased, he succeeded in his petition in the Court of Requests.[410] Similar confrontations arose with increasing frequency, at Weeting in 1604,[411] at Foulden in 1620,[412] and at Bradwell in 1622.[413] At Narford, weld was sown in the fields, with the alleged aim of defeating the fold-course for a thousand sheep, even though weld could well be fed by sheep in its first year.[414] In these and similar disputes, as at Anmer[415] and Bircham Tofts,[416] it seems the tenants deliberately sowed out of course in order to provoke common-law actions that might defeat the fold-course. When they did this, the flock-master usually fed the crop off with his sheep. Then, faced with actions of trespass, he commenced a suit in either Requests or Chancery. The lord of one fold-course, at least, expressly instructed his shepherd to feed off any corn the tenants sowed out of course for the purpose of drawing on common-law actions.[417] On the other side, however, it was sometimes alleged, as at Harthill in 1611, that the lord's fold had stayed in the same field for several years in succession, with the object of defeating the field-course.[418] The conflict between tenants and flock-masters and field-courses and fold-courses thus became also, in some slight measure, a conflict between common law and equity.

Fold-courses impeded improvement in field-courses and stopped the introduction of fallow and smothering crops. Field-courses could hardly be improved without endangering the sheep on whose fold cereal cultivation largely depended. This problem was for long an intractable one, but a solution was eventually provided by a technical innovation, by the introduction of turnip cultivation in farm fields. Turnips gave the sheep much better feed than they could ever have got in summerleys and shacks, and so opened the way to general enclosure and the extinction both of fold-courses and of rights of common, to the common advantage of all.[419]

CHAPTER 5

Town government

All English townships governed themselves by means of by-laws of their own choice, provided only they were not in contravention of the law of the realm. Cities like Coventry, Lincoln and Chester, and many free boroughs like Leicester and Marlborough, as well as most manorial boroughs, had common meadows and pastures, and sometimes common arable fields, and regulated them themselves.[420] Manorial boroughs and lesser townships, lacking independent courts of their own, sought sanction for their common-field by-laws in the courts of lords of manors, normally in the courts baron or customary. But the by-laws so sanctioned in these courts emanated from town meetings. Although servants and others with no stake in the land were not entitled to vote, town government was fundamentally democratic, all matters being decided by a consensus or a clear majority of all the heads of houses of husbandry, their tenants and subtenants, whether men or women. This form of local government was already well established in the thirteenth century and had apparently originated in Anglo-Saxon times.[421]

The by-laws, plebiscites, orders, ordinances or statutes, as they were variously called, were proposed by the by-lawmen or by general town committees, sometimes aided by subcommittees, to the town meetings, which were usually biannual, being held in the spring and the autumn. It was in these meetings, too, that the by-lawmen were elected and shepherds and other town servants designated. The meetings normally immediately preceded the court baron, where the by-lawmen might well be on the jury and where the homage almost exactly coincided with the membership of the town. Where the manor included several townships, one or more of these might be subject to a submanor, and then the submanorial business was normally transacted in the courts of the head manor. Where the by-laws were to be applicable to two or more of these townships, they would meet together to agree joint ordinances. Some townships, however, were divided between two or more manors, and then it was usual to seek sanction for the by-laws in all their courts.

In a court baron, the draft by-laws were presented by the jurymen empanelled from among the suitors. If the laws received the consent of most of the homage and the assent of the lord, they were ratified and promulgated in open court, and, in late medieval and early modern times, were then usually, though not always, recorded in the court books and rolls; and a draft or copy of them would then be retained in the court files. Then the by-laws would be published, by announcements in church, by being cried by the town crier, and, in later times, perhaps also by being posted in the church porch. Infractions of the by-laws, if proved by evidence to the jury in open court, were punished by monetary penalties ('pains'), forfeited either to the lord of the manor to help defray the cost of holding court, or half to the lord and half to the informer, or, in part at least, to the township, on its own general or special account, or to the church, or, in the sixteenth century and after, to the relief of the poor. Without these sanctions, township government would either have broken down or been conducted with grave difficulty. If he cultivated his own land himself, the lord had a say in the by-laws just like the next man; and the lord, or his steward, often took an active part in framing some of the orders. Anyway, the lord had to give his assent to all ordinances before they became valid; but he would only refuse assent if he believed a by-law infringed his own rights, and almost always granted it freely, for good laws helped to keep his property in good order.[422] The lord himself might occasionally be ordered in court to conform to an ordinance, or this be made provisional on his compliance, or he be fined for non-compliance, for he was no more above the customary law than the king was above the law of the realm.[423] He stood in much the same relation to the court as a constitutional monarch to his parliament.

That the by-laws were made by common consent or majority voice is everywhere apparent. Orders, pains and appointments are 'agreed upon by the jury then and all the assemblie there', 'made by the whole jury with the consent of the whole inhabitantes', 'ordeyned by the holl courte', 'upon full and mature deliberacion of all the tenants', 'uppon longe consideracion and debate had amongst all the tenantes'. Such phrases recur at every turn.[424] A court record recites how 'The honorable Henrie Hastings esquire

Lord of the same manor beinge present with the tenants of the same and a great debate beinge theare had and questions moved by some of the tenants ... and theare heard att large. And thereupon by full consent yt was ordered ... and agreed ...' Not all 'orders propounded to be considered of by the tenants' were agreed to; some were rejected and the bills struck through; and that all were open to argument is shown by the preambles advanced to advocate them.[425] Occasionally it might be thought necessary or advantageous to secure a clearer measure of agreement than could be demonstrated by a mere show of hands. Then it would be 'by the allowance of the ... lord ... and with the good liking of the wholle homage' that 'there was orders sett downe and agreed upon and subscribed unto by the homage and inhabitants of the town.'[426] Sometimes it was necessary to petition the lord for his assent to orders that might possibly infringe his own rights. An obnoxious coney warren is ordered to be put down only 'upon a peticon in wrytinge under the handes or sheepe brandes of the ... tenantes'.[427] If the demesne farmer proved refractory, it was perhaps best to appeal to the lord or his estate officers directly, as in the following example:

The freeholdderes and tenents of Collingborne Duses most umbel wise complayning unto your worshepes of diferes and sundri grifes dune by Mr Gefferres. He doth eare up a pease of common called Wigeli contrari to costom and rite. He dide also doe us rong in setting his fold at Bearescombe before his dae of costom came; aulso he doth an ingre in throwing stra in the comming pond; oulso he doth eare up the summerfeld contrari to custom.[428]

Orders made in courts baron were not always recorded, even in later times, for the townsfolk might not think it worth their while to pay a fee to have them entered in the court rolls. Nevertheless, from 1277 onwards, if not before, it was usual to record all but the most routine of by-laws, and agreements and orders entailing the special assent of the lord were almost invariably entered in the court rolls. Where an agreement was concluded outside the court, the whole homage might request its enrolment, asking 'that the said agreement and ordinance be brought into the court and the steward ... cause the same to be inrolled and ingrossed in the court rolles there to remayne matter of record'.[429] Thus there was an agreement for floating watermeadows, which said agreement all the said parties at this courte desired to have entred in the rolles of the courte of this mannor and that thereupon an order should be made for the byndinge all the said parties to performe this agreement upon paynes and penalties to be therein expressed, being a busines conceived to be very behoofefull and beneficiall to all the inhabitants of this mannor.[430]

Common fields, it must be emphasised, were in many respects under common management. Field-courses and pasturage stints decided by town governments were enforced by manorial courts, or, in default, by some specially contrived legal device. In the interests of property, unauthorised and injurious gleaning and gathering were forbidden. In the public interest, commons were protected from depredations. In the interest of landlords and tenants alike, off-farm sales of straw and manure were prohibited. To save the good farmer from the bad, the extirpation of thistles and other weeds was compelled. To preserve grassland from destruction, pigs' snouts had to be ringed, and, to safeguard hedges, people and beasts alike, cows' horns knobbed. To protect the line of great horses against degeneration, down breeding on the commons was prevented by the exclusion of inferior animals. To guard common flocks and herds from infection, horses with mange or fashions, distempered cattle and rotten sheep were forbidden the commons. Animals dying of infectious diseases had to be buried well out of the way. Unless the duty fell on the parson or one of the freeholders, the township took on itself the supply of common bulls and boars. Each common flock had its complement of good and sufficient public rams, to the exclusion of all ridgels and perhaps of all rams of less than a specified monetary value.[431] Common fields were not much infested with hedgerow birds,[432] and crows, sparrows, moles and other vermin survived only despite the measures, orders and rewards for their extermination. Crows' nests were to be plucked down and crow nets were provided at public expense. Molehills had to be spuddled. Mole and crow catchers and scarecrows were public employees,[433] just as were common shepherds and other herdsmen.[434] Township governments often provided such services and facilities as pond-making, ditching and draining, floating, planting and plashing quicksets about the common closes and fields, buying hay for common flocks, mowing thistles, cutting thorns, and knobbing cows.[435] Later, when common grasslands came to be sown down with sainfoin or clover and 'seeds', these were often purchased and sown by the town officers. In some townships, too, turnip fallows were managed in common, the roots being hurdled off in much the same way that meadows were allotted by the acre or the swathe.[436]

Even where common fields were not concerned, town governments busied themselves with orders for such things as hedging, ditching, the ringing of pigs, the extermination of weeds and vermin, and the prohibition of outside sales of straw and manure.[437] Town meetings and court orders were by no means confined to agricultural matters; they concerned also such things as the provision of a common grindstone,[438] the supply of drinking water (either by water-carriers with carts or leathern budgets, or by wells, or by conduits and taps),[439] the appointment and payment (sometimes in common rights) of schoolmasters,[440] the repair of roads and embankments,[441] the prevention of nuisances[442] and of fires caused by faulty chimneys and pipe-smoking near ricks and stacks,[443] the control of inmates and undertenants,[444] the provision of poor relief in various forms,[445] and the regulation of servants and labourers.[446]

The increasing volume and complexity of township business often demanded the services of a salaried town clerk,[447] who also took charge of the town archives, which were kept in the common chest and lodged in the parsonage, the church tower, or some other place of safety, and to which both the town clerk and the lord's steward had free access.[448] Some of the larger and more go-ahead townships built their own town halls, but many continued to meet in inns or in school and church-houses, more particularly in the vestries.[449] The Glatton town meetings were held in the church-house. The inhabitants were summoned by the ringing of the great bell 'twice or thrice' and were bidden to stay 'until the matters of chief interest to the aforesaid town shall be ordered and determined'. The tolling of the bell was the signal sounded by the Hillmorton malcontents to call the townsfolk together to 'make a common purse among them' by imposing on the whole township a rate of 2*s* 6*d* a farthingland or cottage, in order to oppose the lady of the manor in an enclosure dispute.[450]

Considering all the business they undertook, it is not surprising that township governments were often employers on some scale, with whole or part-time staffs ranging from clerk to dike-reeve or floater or to crow-keeper or water-carrier. These were all paid, and public works and business financed, by rates levied on the all the inhabitants, that is on all persons with property in the township that could be distrained upon in the event of non-payment. Thus menial servants and other propertyless people, not being classed as inhabitants, were exempt from paying rates, while absentee proprietors were due to pay them, for they were able to inhabit whenever they chose, and were deemed inhabitants. This meant some men owed rates in two or more townships. Each rate was laid separately and its proceeds earmarked for a particular purpose, so that, justly enough, people only paid for services they themselves enjoyed and subscribed to, and were not obliged, for example, to contribute to the shepherd's wages unless they put sheep in his charge. Non-payment of an applicable rate, however, could be followed by distraint. Town rates were levied either by the pound (or poor) rate or by an acre rate. Poor and pound rates were often nominally one and the same, for all poor rates were by the pound, and pound rates not levied especially for the poor were nevertheless, on this account, often called poor rates. Pound or poor rates were gathered according to the assessed values of the premises of each household and their proceeds were employed for any purpose that concerned the whole township and all its members. Acre rates were intended to be applied to specifically agricultural services and were therefore levied by the acreage of the agricultural land held.[451] But a pound rate might well be used to wage a molecatcher, for moles were everyone's concern, harming as they did the cottage garden as much as the demesne farm.[452] In addition to the rates, some townships' funds were supplemented by common properties or interests like village greens and some of the meres and headlands in the common fields.[453]

A minority of the townsfolk, consisting of a few gentlemen and the more substantial of the yeomen, usually filled the major offices and took the lead in town affairs. They were the 'wise persons' or 'discreet men' who proposed the stints for the common pastures and to whom the herdsmen rendered account of the farm animals on the commons and reported those that were diseased. Without first informing one of the officers, no one might let his common rights to another. These wise men also decided when the field was ready to be commoned. Assembling farmers, servants, cottagers and labourers, they supervised trenching and draining. They framed the bills or draft by-laws, and it was in their discretion to relax statutes should they prove temporarily inconvenient by reason of the vagaries of the weather or other immediate circumstances.

They saw to buying hay for the common flock and arranged the hitching field. They were in receipt of the rates and took charge of the common purse, rendering accounts at the expiry of their terms of office. The selfsame men, many of them, together with others of their kind, were to the forefront in all that concerned the townsfolk. As jurymen, they were judges as to fact in the manor courts. As the town's discreet men, they often acted as arbitrators in minor disputes. As overseers of the poor, they gave alms to the needy and, subject to the consent of the lord of the manor, authorised the erection on the waste of cottages for the poor, pest-houses and similar buildings. As constables and deputy constables, they often helped to keep the king's peace.[454] In short, as always in all walks of life, leadership came from the few and business tended to be confided to those best able to deal with it.

This system of government could not obliterate human imperfections. Occasionally wealthy farmers usurped the stints allotted to the less fortunate or otherwise surcharged the commons, refused to present one another in court, and conspired to crush anyone who offered to present them, so that only the intervention of the lord's steward of courts could prevent the corruption of government.[455] Occasionally, too, it might happen that some of the discreet men, in their capacity of overseers of the poor, were accused of partiality in the distribution of relief.[456] Yet when there was no body of substantial tenants to be the town's discreet men, government suffered. So many of the best houses in Cerne are now 'fraught with beggers', reports John Norden, that the town is

> most unorderlie governed and as uncivillie as if there were noe magistrates, for the officers are weake men, for they that injoye the principall howses of the towne dwell from them and let them to a manie of base people, meere mendicantes, as namelie Mr John Notley hath a fayre howse but noe competent undertenant, Mr Lovelace also a fayre howse full of poore people, one Mr Fowell, Mr Devenish and one John Williams, which last hath a fayre howse and hath put nere a dozen lowsey people in it.[457]

Either extreme, excessive power or baseness, might lead to dissensions among the jurymen,[458] to the steward charging jurors with perjury in making untrue presentments,[459] or to gross contempts of court.[460]

It is clear that courts baron, like the complementary courts leet that dealt with minor cases in law, were fair, swift, convenient, cheap and efficient.[461] In them, each township had its land registry, its law courts, and its law enforcement, all run by the townsfolk themselves, with only the assistance of a steward learned in the customary law and the law of the realm. For this reason, manorial courts survived customary tenures in some places and continued generally until the end of the eighteenth century, sometimes longer.[462] And when finally they came to an end, the judicious opinion was that

> The revival of manor courts, throughout the kingdom, (or the establishment of other rustic tribunals of a similar nature), could not fail of producing the happiest effects. They are the most natural guardians of the rights of villagers, and the most prompt and efficient police of country parishes.[463]

In many enclosed and some unenclosed townships, the manors were dismembered long before this time, with the result that town governments lacked sanction. Then, as John Aubrey says, 'the meane people live lawlesse'.[464] An insidious attempt to deprive existing manors of sanction and authority had been foiled by Queen Elizabeth when she refused her assent to a Bill to hand over all responsibility for crow-catching to churchwardens. In 1662 a Bill introduced the year before 'for making orders and bye-laws for well ordering and governing of common fields' had a third reading, but was not enacted. It was not until 1773 that Parliament allowed elected field-reeves themselves to enforce by-laws passed by town meetings, if manor courts were no longer held.[465]

In the meantime, when there was no court baron to enforce the by-laws for husbandry, resort might be had to the hundred court, but, as the people of Bledington discovered in 1647, a hundred court would enforce the payment of penalties for breaches of these by-laws, except perhaps for a short interim, only if the full agreement of all the townsfolk to the ordinances could be proved beyond doubt, and such agreement was not always easy to obtain or prove. The remedy usually adopted in practice was to call a special town meeting to draw up indentured articles of agreement, so that compliance with field orders could be enforced by the taking of bonds liable to forfeiture. Officers could then be appointed and pains and penalties imposed, the entire proceeds going to the township and being devoted to the poor or to the maintenance and repair of the church-house. But even this expedient depended on the unanimity of the townsfolk or the smallness and weakness of any opposition.[466]

The eventual decline and supersession of the manor hastened the confusion of township and parish. Since jurors, discreet men, overseers, constables and churchwardens were usually drawn from the same group of men, their persons were often identical, and this led to some confusion of functions. Foremen of juries are found calling vestry meetings to discuss clover cultivation, churchwardens receiving the proceeds of the sale of common grass and supervising the sheep commons, and churchwardens, constables and overseers of the poor all being ordered in the court baron to present their accounts by the next court day.[467] The distinction between township and parish had always been somewhat blurred, for, except in mountainous and other parts where settlement was scattered and several townships, tithings or hamlets went to make a parish, the parish was constructed on the existing bounds of the township and the two often coincided geographically. Some Acts of Parliament were worded to apply to the township or the parish, for the town meeting was often in the vestry and the relief of the poor was related also to church business.[468] When manor courts were extinguished, the distinction between town and parish meetings became even more blurred and finally disappeared, so that the civil parish could be defined as a place for which a separate poor rate was or could be made and for which a separate overseer of the poor was or could be appointed. In this way it often happened that several civil parishes were carved out of one ecclesiastical one.[469] What was until recently called a parish council was essentially the old town government, while the old parish council is now called the parochial church council.

CHAPTER 6

Common field and severalty

There have long been two schools of thought on the merits of common fields. One school has romanticised them as a Garden of Eden unfortunately disrupted by the serpent of commercialism or a land of Cockaigne wickedly turned into barren sheepwalk by grasping landlords. The other school has made them out to be sloughs of sloth and ignorance and a 'miracle of squalid perversity'.[470] Both are wrong and one-sided.

As has so often been said, the lack of hedges was a disadvantage in many common fields, for it meant a shortage of wood for fuel; but it was also some advantage, for there was less harbour for birds and vermin and less danger of the mildews that frequently injured corn crops in deep, enclosed countries, especially if the hedges were not properly slashed and plashed.[471] Common fields can hardly have fostered as many weeds as present-day hedges, hedgerows and road verges. It has been repeated ad nauseam that the absence of hedges and mounds invited intrusions and encroachments, especially in ploughing; but such trespasses were far rarer than exaggerated strictures may lead one to suppose. More disputes arose, perhaps, from adjacent townships intercommoning the wastes. This was one reason why many of these townships agreed to divide the commons between them and to enclose one against the other.[472] What, in fact, is more remarkable than disputes occasioned by common fields is the highly successful co-operation achieved in, among other things, the management of common flocks and folds, eliciting William Marshall's observation:

> Such is the power of invention, when urged by natural necessity, that even the lowest class of farmers are enabled to keep sheep and fold their arable lands with a degree of propriety . . . Theory may suggest that endless difficulties, and disputes, must necessarily arise, from individual properties, and separate interests being intermixed and rendered common. But the long established practice, under notice, serves to show that, where a common compact is requisite, to serve the interests of individuals, men's

minds, seeing the reason and fitness of the regulation, become reconciled to small difficulties, and are satisfied to give and receive, reciprocally, as circumstances require.[473]

That livestock infections were endemic in the commons has never been demonstrated and is difficult to credit in the face of the relevant by-laws. It cannot be true to say that 'sheep could not be herded with success in open commons, still less on the arable land of village farms, and small holdings were incompatible with large flocks'. To write that common-field farmers 'commanded little or no manure for their arable land and were practically dependent on the sheep for fertilizing the soil' is merely to say they had nothing but the best. It may well be that some common-field sheep in earlier times were reduced to the lowest possible level in winter and barely survived on straw and tree-loppings,[474] but no sheep lived harder in winter than the Romney Marsh ones, which never saw a common field.[475]

The allegedly excessive cost of tilling dispersed parcels in common fields is exaggerated. All ploughing in the ridge-and-furrow system, which was the one by far most widely used, involved tilling discrete pieces, no less in enclosures than elsewhere. No one has yet shown how riding a string of horses out to a furlong where the plough had been left overnight and there ploughing some ridges, was more costly than comparable work in closes. Ploughing costs became exorbitant only when the parcels were excessively fragmented or the fields themselves too large and parts of them too remote from the farmsteads.[476] The simple remedies applied were, respectively, the consolidation of small common-field parcels into larger ones,[477] and the division of the township into two towns, tithings or 'ends', each with its own fields.[478]

Common fields are often deemed to have been inefficient because bare fallows were made in them. But fallow crops were grown in the hitching fields and later in all or most of many common fields. Moreover, bare fallows are not always bad practice. In much every year's ground, where they were often avoided, it was by no means unknown for the land to became choked with weeds. Even in enclosed fields some bare fallowing and other cleaning cultivations had to be undertaken from time to time. It is wrong to suppose bare fallows were found only in common fields or fallow crops only in severalty. Nevertheless, in common fields, bare fallows were frequently preferred to turnip fallows and short clover leys, even where both were practicable. This was because the cultivators' paramount interests were in growing bread and drink corn for domestic consumption. These men needed above all regular, even if moderate, yields of good wheat, barley and other corn, and the likeliest way to get these was to precede the crops by bare fallowing for twelve or eighteen months. Turnip fallows and clover leys could not be tolerated if they diminished the chances of good barley and wheat. Moreover, it so happened that where the common-field farmers tried and rejected turnips and clover, these crops were almost equally unsuited to enclosed severalty.[479] For the objects common-field cultivators had in mind, then, bare fallows were frequently better than fallow crops or clover. Bare fallows allowed them in large measure to avert the risk of being left without bread and dying of starvation. But they also largely averted this risk by dispersing their parcels of arable between lowland and upland, heavier and lighter soils, nearer and farther locations, sunnier and shadier situations, and so on and so forth, because weather that spoiled crops in some lands was less harmful in others. But this was not strictly a peculiarity of common fields; where advantageous, similar dispersals were made or continued in enclosed townships.[480]

Many of the comparisons made between common fields and severalty are misleading. For example, what Thomas Tusser was really contrasting was not champion and enclosed farms in general, but farming in the Norfolk Heathlands and in High Suffolk. What he most disliked in the Norfolk Heathlands were the inconveniences of the fold-course to small tenant farmers and the disputes that arose between them and the fold-masters, circumstances that were peculiar not to common fields in general but to this country in particular. An additional, and purely fortuitous, reason why Tusser so disliked his sojourn in West Dereham was that he was adversely affected by a boundary dispute between his own and a neighbouring lord:

> Then did I dwell in Dirham sell,
> A place for wood, that trimlie stood,
>   With flesh and fish, as heart would wish:
>   but when I spide
> That lord and lord could not accord,
> But now pound he, and now pound we,
> Then left I all, bicause such brall,
>   I list not bide.

Another comparison often cited is that in which Fitzherbert explains how to make a township of twenty marks a year worth twenty pounds. This has the merit of relating to one and the same Midland township. His scheme is to enclose the common fields and pastures by agreement and convert permanent tillage and permanent grass to up-and-down land. This would certainly have raised the rental value by one third, but not by virtue merely of the extinction of common rights, for the improvement came mainly from the heightening of productivity by the adoption of convertible husbandry. The comparison is thus rather between two agricultural methods than between common field and severalty. Similar improvement in rental values could have been, and often was, attained on land always in severalty, simply by replacing permanent tillage and permanent grass with up-and-down husbandry. Fitzherbert rightly assumed convertible husbandry was feasible only in the absence of common rights, but it does not follow that all extinctions of common rights necessarily led to such technical change and improvements in the value of the land.[481]

Comparisons were frequently made between the rental values of common field and severalty; but all too often fields with different soils were contrasted, as thin-soiled common fields and deep-soiled closes. Such comparisons carry no weight.[482] Rental values sometimes trebled after enclosure, as they did in the Cotswold Country, but not merely on account of the enclosure itself. Turnips or cultivated grasses were either introduced for the first time or grown on a much greater scale than before, and improvements in rents came from this rather than from the extinction of common rights. Division and enclosure in the plain countries led to great rent increases, just as Fitzherbert said they would; but this was due less to the ending of common rights and more to the laying down to grass of old tillage and the ploughing up of old grasslands.[483] The rents of enclosed farms were often double those of common-field ones; but then the latter were tithable and the former tithe-free, i.e. either the tithe was commuted into money and paid as part of the rent, or an easy *modus decimandi* had been arrived at.[484]. Moreover, the enclosed farms were usually larger than the common-field ones they replaced and were better able to take advantage of some economies of scale. For one thing, all small arable farms, whether in common field or severalty, needed more men and horses an acre than did large ones.[485] All the same, experts held that enclosed land in severalty was worth at least a third more than, and up to twice as much as, comparable open common field, except where this was every year's land.[486] The mature conclusion formed by William Marshall, who was without peer as an agricultural expert and observer and a strong advocate of enclosure and the extinction of common rights, was that

> Open lands, though wholly appropriated, and lying well together, are of much less value, except for sheep-walk or a rabbit warren, than the same land would be in a state of suitable inclosure. If they are disjointed and intermixt in a state of common field or common meadow, their value may be reduced one third. If the common fields and meadows are what is termed Lammas land, and become common as soon as the crops are off, the depression of value may be set down at one half of what they would be worth in well-fenced inclosures and unincumbered with that ancient custom.[487]

The crux of the matter is that common fields always remained true to their origins in being worked mainly by small cultivators in family, part-time or dwarf farms with relatively low outputs and high costs. Nearly all occupiers of common fields, including those with non-agricultural interests also, were, as cultivators, concerned first and foremost with mere subsistence. They had to sell some goods or services in order to meet their rents, rates, taxes and other unavoidable outgoings; but the great object of all their agricultural endeavours was to ensure that they and their families were regularly supplied with bread and drink corn and, if possible, with some butcher's meat and dairy produce. These, together with their kitchen gardens and orchards, gave what was needed to sustain life, provided only that they gave it regularly, year in and year out. The great risk to be averted at all costs was harvest failure and starvation. As for risky ventures in search of high monetary returns, they were studiously avoided if they cast a shadow of risk upon mere subsistence, upon life itself.

The respective merits of common field and severalty were these: severalty was better for farming for the maximisation of monetary profits, common field for subsistence; severalty for capital farmers and their landlords, common field for family and part-time farmers in cereal-growing countries. As for technical standards, individual initiative being the mother of improvement, invention and innovation, as well as of failure and disaster, it has truly been said, 'Severalty makes a good farmer better and a bad one worse'.[488]

CHAPTER 7

Common fields overseas

We may learn something from a glance at common fields in Europe and other foreign parts. We are primarily concerned with the plains of north-western Europe, but shall cast our net wide.

Southern Europe we shall take to mean the part made up exclusively of the following: the Mediterranean region; all France south of an imaginary, irregular and uncertain line running east-southeast from the east coast of the Cherbourg peninsula and skirting the south of Normandy and the north of the Touraine; the vale of the Moselle and the rift valley of the upper Rhine, both extraordinarily sunny, mild, fertile, and withal thoroughly Romanised; and all that lies south of the Rhine and the Carpathians. In this region, the climate, generally speaking, the mountains apart, is warmer than the British in all seasons and markedly drier in late spring and summer, so discouraging the cultivation of grass and spring-sown cereals. This was and is a world of wine and olive oil, unlike the northern one of beer and butter. In the more westerly parts of southern Europe, common fields have long been cultivated, common of shack enjoyed, and mutually agreed field-courses followed. Usually the land has always been tilled with the *aratrum, araire* or drag-plough that only scuffles up the surface of the soil and, by dint of repeated ploughing and cross-ploughing, pulverises it, so conserving the underlying moisture. Since cross-ploughing was the ordinary practice, the fields furlongs and parcels tended to be squarish. Southern Europe has many and wide areas of thin, poor, and infertile soils, like those of the Alps, the Apennines, the Massif Central, Brittany, Maine and Bearn, and here tillage has been mostly temporary and shifting.[489] But in the fertile soils, two- and three-field courses have been used, since ancient times in the west and since ever more recent ones as one goes eastwards. The most usual course has been (1) crop, (2) fallow, within which four-crop rotations have sometimes been accommodated. Many places have followed three-field courses, consisting nearly always not of (1) winter crops,

(2) spring crops, (3) fallow, but of (1) crop, (2) fallow, (3) fallow, as alluded to by Virgil when he wrote, *illa seges demum votis respondet avari agricolae, bis quae solem, bis frigora sensit* – 'The greedy husbandman likes best that mould that twice hath felt the sun and twice the cold.' Two-year fallows have permitted more podware and other fallow or smothering crops, especially long-standing ones like flax, hemp or cotton. In both two- and three-field courses, much of the fallow has been sown to rape, sesame or sorghum, or to turnips, radishes, peas and beans for the table, or to vetches and other hitch crops for riding and carriage horses. In modern times, too, maize has been much grown in the fallows. And everywhere the vine has been assiduously cultivated.[490]

The poor, thin soils of Norway, Sweden and Finland formed another region where the ard or scratch plough was supreme. Here the arable consisted mainly of shifting and temporary tillage in which oats were grown year in and year out until the soil was exhausted and abandoned. The land was often cleared by burning off the wood and scrub and then sowing seed in the ashes, giving great crops in the first year and diminishing ones thereafter. By the eighteenth century, permanent cultivation had developed in Scania and a few other favoured parts. Here two-, three-, and even a few four-field courses were employed, but some of the two-field ones excluded winter corn.[491]

We may also cast an eye much further afield. Although in the many hilly and wooded parts of India slash-and-burn and other kinds of shifting cultivation are the general rule,[492] in wheat-growing regions like the Punjab the two- and three-field courses are in permanent common fields not much unlike those in northern Europe.[493] China and Japan, too, have had common fields with dispersed lots and parcels.[494] As for the rest of Asia (paddy-fields excepted), and much of Africa and America, temporary, shifting cultivations are still widespread.[495]

Our interest and attention is thus focused on the plains and tablelands that stretch across the main body of northern Europe. This is where ridge-and-furrow ploughing, permanent cultivation and common fields became general enough to allow comparisons and contrasts with England.

Even in this region, temporary and shifting cultivation, sometimes combined with perpetually cropped infields, never died out entirely. Under various, and often highly descriptive, names it long continued in hillier districts, wherever the soils were thin and poor, and on better land in backward and remote parts, just as it did in England. Temporary cultivations often took the form of *Brennwirtschaft* or beat-burning, or of *Wildfeldgraswirtschaft*, where rough pasture was ploughed up for a few crops and then left to recover of itself. Just as in England, too, these various shifting cultivations often took the form of temporary common fields or *Vöhnden*,[496] and then, just as in England (and India), their parcels were allotted to the townsmen in direct proportion to their rights of common of pasture, which were themselves often proportionate to their holdings of permanent cultivation. Allotment itself proceeded either by casting lots, choosing straws, or some such, or by following a customary rotation of allotment originally based on some such procedure. 'Lot is acre's mother.'[497]

In later times, mostly in sub-Alpine districts and in Denmark and Holstein, much of the land was in up-and-down husbandry in severalty – *Leyfarming, Egarten, Koppel, Wechsel* or *Neufeldgraswirtschaft*.[498] Generally, however, most of the arable became what it still largely is: permanent tillage in common fields. Such fields once obtained even in the Friesian islands and in *het Gooi* near Amsterdam.[499]

In trying to satisfy his curiosity about the history of European common fields, a mere Englishman encounters unexpected difficulties, for in England common fields have long been rare vestiges of a bygone age and a fit subject for historical enquiry, whereas the majority of Europeans take them for granted, as though they were features of the landscape as eternal and immutable as the very lakes and mountains. Not knowing anything but common fields, they have failed to find much significance in them. Partly for this reason, European historians have mostly taken little notice of them and have left their study largely to geographers, whose interest has naturally been directed chiefly to the pattern of settlement and to the shapes and outward forms of townships and fields, and, until recently at least, hardly at all to their function and operation. Their studies are by no means without interest and we shall have occasion to refer to them later; but for the time being it will suffice to say that, as in England, the shape and form of townships, fields and furlongs depended to a great extent on the nature of the terrain they occupied. Where, for example, the land was all much of a muchness, townships tended to be compact and roundish or squarish, while elsewhere, where markedly different kinds of land were found in close propinquity, townships tended to take the form of strips of territory running across all the different kinds of land, as from upland towns to the heart of their associated marshlands, from valley bottoms up to the summits of downs, hills and mountains, and so on, and so forth, for lands that were cultivated and managed in conjunction with one another had likewise to be united in holdings and townships. Alpine townships run from the river and its valley up into the *Voralp* and the *Alp*. Many Indian townships extend lengthways from the dry-season water level, across the flood-line of the rainy season, and then on through the uplands. As to the regularity or irregularity of the ways in which arable land was divided up into fields and furlongs and distributed in dispersed parcels, this, as in England, depended largely on whether the settlements had been planned and organised by a single lord, authority or association, or had been developed independently by a number of individual families. As for the forms of fields, in their German nomenclature, the *Blockgemengflur mit blockartige Flurverteilung* (field-system composed of intermixed blocks of land with subdivisions conforming to the block principle), or some parallel system conditioned in its forms by a different type of terrain and settlement, may perhaps be taken as an early step on the way towards a common field; but only where each furlong has been divided up into parcels, by *Parzellierung* (parcelling out) and *Vergewannung* (formation of furlongs composed of such parcels), can a true *streifige Kernflur* or *Gewannflur* (rudimentary or complex common field), be said to exist.[500]

Another great interest of European geographers and other scholars has been the counting of the number of topographical fields into which a township's lands have been divided, and from this number the divination of the field-course employed, so that it may or not be classified as a *Zwei- oder Dreifelderwirtschaft* (a two- or three-field system). Here we are instantly plunged into dangerous waters, for, as in England, the number of fields gives no good guide to the field course or courses employed, for the fields themselves were often divided or grouped to form shifts for the field-courses.[501] Moreover, Europeans generally have become accustomed to thinking in terms of generic distinction between, on the one hand, *Dreifelderwirtschaft* (with two- and four-fold courses being taken, not unreasonably, as mere variants), and on the other, all other forms and states of arable, as though all lands in the three-field system were necessarily in permanent tillage, and as though the *Dreifelderwirtschaft* were a set and uniform system.[502] In reality, not all *Gewannfluren* or lands in *Dreifelderwirtschaft* were even in permanent cultivation. As in favoured parts of the North-eastern Lowlands of England,[503] three-field courses, in many places in Russia, were practised in strictly temporary cultivations. After an unbroken run of the three-field course repeated three or four times, the whole field was abandoned, the total of abandoned land being usually much more extensive than the tillage.[504] Near the Hunsrück, a three-field course was followed in temporary tillage plots.[505] The most usual course in the Drenthe common fields was a two-field one of (1) winter rye, (2) spring rye, but large parts were abandoned from time to time and allowed to revert to waste (*woeste*), so the field itself was largely not in permanent cultivation. Similarly, in Maine, even in the nineteenth century, the two-field courses ran for six years, after which the land was abandoned for three years. In Brittany two and three-field courses were followed by abandonment for six years or more. Much the same practice obtained in la Beauce in the thirteenth century.[506] Incidentally, a two-field course is employed in shifting cultivations in the arid lands of modern Syria, there being much more land abandoned than actually tilled, so that 'la mise en valeur n'est pour ainsi dire qu'un bref épisode entre les longues périodes de vacances'.[507] Similarly, in eighteenth-century Norway, in the leveller parts of Ostland and in Trondlag, a four-field course was followed, but every third year alternating portions of the tillage were abandoned.[508]

The short and long of it is, almost any type of field-course may be practised in almost any type of field, there being no causal connection between the type of field and the type of field-course. An undivided field, without separate furlongs, may be suitable for temporary, shifting cultivation, and a group of fields, each made up of several different furlongs, with each furlong split into many parcels, for a three-field course;[509] but such fields are not necessarily cultivated in these ways.

Moreover, there were many more kinds of field-course than are dreamed of by some writers. Not all two-field courses, for example, included a fallow year. One, practised in Drenthe, the Veluwe, Lower Saxony and thereabouts, consisted of (1) winter rye, (2) spring rye; all concentrated in one small area of permanent tillage enriched with a compost of farmyard manure and of turves skimmed from the spacious heaths. This could almost as well be described as a one-field course in monoculture. The same might be said of the four-field variant where three crops of winter rye were followed by one of spring.[510] When we are told that this or that township practised a *Zweifelderwirtschaft* or a *Dreifelderwirtschaft*, we are not really very much further forward, for we are still left in ignorance as to what exactly the field-course was and whether it was being followed in permanent or temporary cultivation. Still less can we be sure whether the fields were common or not. Many Europeans, Germans especially, seem to take it for granted that the two- or three-field system betokens common fields throughout the township, which is understandable, for west German demesne lands were often intermixed with those of the tenantry.[511] But commonage cannot be deduced from the two- or three-field system or from a two- or three-field course.

Largely divined as it is from modern estate plans and maps, the picture that emerges is indistinct, ill-defined and impressionistic. Nevertheless, standing back from it, we can dimly make out a region where *Gewannfluren* and the *Dreifelderwirtschaft*, in its loosest and most general sense, roughly coincide, and this region, we know, was at one time, and still largely is, in permanent common fields, often with three-field courses. It extends across northern France and most of northern Germany, into Denmark and into northern central and eastern Europe, including most of Poland, Bohemia, Austria, Latvia, Estonia, Lithuania and much of the Russian empire.[512]

As to exactly what kinds of field-courses were practised in this region, we are still left rather in the dark. To our initial surprise, we are told only of one kind of three-field course, that followed in the downlands of England, viz. (1) winter corn, (2) spring corn, (3) fallow.[513] This sets us wondering how deeply the courses themselves have been investigated and how far the sole three-field course known to the writers of text-books has simply been assumed. Recently, however, some Continental scholars have started to show a renewed interest in the nature of field-courses and have turned their attention away from the *Dreifelderwirtschaft* to the *Dreizelgenbrachwirtschaft*, from the three-field system simply conceived to the three-field course.[514] We now learn that one three-field course found in Lower Saxony, Drenthe, Overijssel and Groningen was (1) winter rye, (2) winter rye, (3) fallow, perhaps partly sown to buckwheat or black oats.[515] Nevertheless, it seems a course of (1) winter corn, (2) spring corn, (3) fallow, was indeed the three-field one most widely followed in the Pays de Caux, the plains of Caen, Falaise, Hainaut and Cambrésis, the Ile-de-France, the plateau de Langres, the Birkenfeld district, and probably much of the French and German plain.[516] Two-field courses appear to have been followed in many places,[517] but we learn little of the nature of either these or the four-field ones, except in the Veluwe and Lower Saxony district, and near Zweibrücken, where four-field courses were merely modified two-field ones.[518] We are told more in detail about the lower vale of the Rhine, Cleves and Gelders, and Flanders, where the livestock were house-fed and, as was the usual way in market-gardening in England, agriculture was combined with horticulture, and alternating parts of the tillage were dug with the spade for four, five and six-year rotations of such garden crops as cabbages, turnips and beetroots. Here, 'De Spa is de Goudmijn der Boeren' – 'The spade is the peasant's goldmine'. But precious little of this region lay in common fields.[519]

All the field-courses incorporating fallow years admitted of catch crops on the fallows. Hitching fields were found in many parts, and in the district of the lower Rhine often occupied most or all of the fallow fields, which thus assumed the status of full-blown summer-fields. Much as in England, these hitching and summer fields were sown to lentils, vetches, peas and beans, with oats, millet or spring rye for them to climb by. But there were some points of difference: first, in Flanders and around the lower Rhine, spurry, red clover and lucerne were amongst the fallow crops that were introduced earlier than they were in England; and secondly, whereas in England hitching fields were almost ubiquitous and the horsemeat was destined almost entirely for farmhorses, in Europe these fields were concentrated about the large towns and in the industrial districts, where street and stable manures were plentiful, and the horsemeat went mainly to urban riding, draught and carriage horses.[520] In addition, smothering crops like flax and woad were grown in some fallows, which were then often prolonged into a second year in order to accommodate them.[521] Finally, again as in England, some fields or furlongs were every year's land.[522] And even this sketchy account of field-courses in the European plain simplifies unduly in that two or more different courses were often followed in one and the same township, to suit the nature and accessibility of its various fields.[523]

Generally speaking, however, we are not told much about the field-courses in European common fields in any period before the late eighteenth or early nineteenth century, when the new crops were belatedly introduced into them; and even then we learn more about crop rotations than about field-courses. The diffusion of these crops, which was what occasioned the newly informative nature of the records, was, roughly speaking, from west to east. Sainfoin was not entirely unknown as a crop in the Ile-de-France in 1645, and in the Pays dc Caux, in 1705, an attempt was made, apparently without any lasting success, to grow clover under oats, so as to stand in the succeeding fallow year. By 1763, many bare fallows in Normandy had been suppressed or reduced, as the best farmers took to a course of (1) wheat, (2) clover under oats or barley, (3) clover ley. In the townships with two-field courses, however, progress was much slower, the cultivators fearing for their wheat crops.[524] Eventually, clover and 'seeds' triumphed in the common fields all over Europe. They were introduced in northern France and in Denmark in the late eighteenth century. In the Rhine valley, and, in the nineteenth century, in Bavaria and parts of Normandy also, red clover and *Englisches Raygrass* or *le ray-grass*, meaning English ryegrass, in the Sauerland, clover, tares and other podware, all entered the common fields.[525] By the middle of the nineteenth century, in the Austro-Hungarian empire and the Baltic states, by the early twentieth in the Ukraine, north Russia and thereabouts, similar crops had been taken up in common fields, in such courses as (1) rye undersown with clover, (2) clover mown, (3) clover fed.[526] Similarly, stockfeed potatoes and root crops were adopted by common-field cultivators in Lower Alsace about 1720 and later spread to other parts of western Europe, the two- and three-field courses being adapted for four-, six- and nine-year crop rotations.[527] Later still, similar practices were introduced in eastern Germany and other parts of eastern Europe.[528] From the middle of the eighteenth century, too, maize was grown in the fields of south-eastern Europe.[529]

One feature found in all common fields in Europe, and those in most parts of Asia, was that they depended for fertilisers largely or mostly on animal manure from the stable, cowshed or farmyard or from the sheepfold. In China the want of livestock of almost all kinds led to the large-scale use of human excrement, but this practice was generally eschewed in Europe. Wherever they had enough sheepwalk, the common-field cultivators usually had their common flocks, common folds and common shepherds. In Drenthe, however, the sheep were usually housed at night and their dung carted out; only for a few weeks at the end of summer were they close-folded on the tillage. In parts of Germany and Syria, town flocks were replaced or supplemented by annual visitations from large migratory flocks. In much of Syria the Bedouins' flocks still are yearly folded on the common fields, 'pour la nuit tour à tour sur les champs qu'il désire mettre en culture l'année suivante'. In India and Turkey, goats have often swelled the folding flocks.[530] In much of the Mediterranean world, since dry summers inhibited the growth of grass, few livestock were kept other than for draught purposes, and sheep flocks large enough for folding were something of a rarity. For different reasons, they were equally rare in the Alpine region. In these and similar places, the fertility of the common tillage depended almost exclusively on farmyard manure. And elsewhere, even when sheep were kept in some numbers, they were, in earlier times, usually penned or cotted and hardly used for the fold. Wherever farmyard manure was the sole or chief fertiliser, the difference between land that was mucked (*Dungland*) and land that was not was necessarily very sharp. Only regular mucking could keep land in permanent tillage. Hence, except in the few places favoured with muck galore, the permanent common fields remained relatively small.[531] Where few sheep were kept and folding was not practised, *vaine pâture*, rights of common of pasture in shacks, fallows and elsewhere, could remain more or less unregulated and unstinted.[532] Common herds of cows or goats, as in Alpine countries still, were found almost everywhere, but mainly for convenience and economy of labour in herding.[533]

Everywhere common fields have been accompanied by formal or informal town government. Even in the absence of common fields, such things as common ponds, common woods, common herds, public ways, and the town watch, needed management by a headman, a council of elders, a village meeting or some organised body.[534] In common-field towns, local government attained much greater importance, for the township itself, or its elected leaders, had to decide how new temporary or permanent common fields were to be allotted, what field-courses should be followed, which fields or furlongs should be in which shifts, how the common flock and fold were to be managed, and so on and so forth, and the decisions had often to be formulated as by-laws. Generally speaking, too, the more complicated and varied the field-courses and shifts, the more common management was required, and it was needed most where new crops and new courses were introduced in the eighteenth century and after. The regulation of field-courses and of common of pasture was an inseparable and inevitable concomitant of common-field agriculture. In its most developed form, this regulation or *Flurzwang* (*assolement forcé*) amounted to the policing of the fields and the penalising of all infractions of the by-laws.[535]

But such regulation was not always needed. In India it was hardly found, except for irrigation works. The original agreement on field-courses made when the common field was first instituted, was embodied in the town's custom and usually remained unchanged for ages, obviating the need for by-laws.[536] Before 1945, Romanian peasants co-ordinated all their field-work, even though they had no town governments or similar organisations.[537] Before the policing of fields was introduced in Bavaria, all was arranged and ordered by voluntary, neighbourly agreements,[538] and the same seems to have been true generally: neighbourly agreements preceded *Flurzwang*.[539] Even after all compulsion had been ended in France, the selfsame usages continued as before: 'Toute constrainte a disparu, mais l'usage s'est maintenu.'[540] This spontaneous order, this uniformity of agricultural practices, and this conformity to the custom of the town, it is generally agreed, arose because the presence of fields and furlongs split into small parcels, combined with the operation of regular field-courses, meant that every cultivator in the common fields was compelled by the practicalities of his daily round to co-ordinate his field-work with that of his neighbours. To break the field-course or the shift cycle or to avoid common of pasture, was not merely socially impossible; it was also technically impractical. Wheat sown in the fallow field, for example, could not come to fruition for purely agricultural reasons; and only a fool would have broadcast good seed in such exhausted land. And, of course, if such seed had sprouted, the crop would simply have been eaten off by the flocks and herds. Within the constraints of the field-course, each cultivator was free to choose his own crops and he asked no more. Nearly all common-field cultivators usually wanted nothing better than to go on cultivating in accordance with the customary methods taught them by their fathers. They liked the traditional routines. They liked doing what their neighbours did.[541] And all this accords with all we know about neighbourly agreements, by-laws and manorial courts in England.[542] By-laws, regulation and legal compulsion seem to have been evoked by such things as the introduction, by agreement, of novel field-courses, crops and rotations, or by a venturesome few embarking on some new practice unilaterally, without waiting for agreement. In the long periods between such rare events, however, little or nothing was needed by way of compulsion. And all the regulations and by-laws were, we must remember, the fruit of agreements freely entered into by the vast majority of the cultivators, and coercion, mild as it was, was solely to guard against the possibility, seldom realised, of small recalcitrant minorities.

With the passage of time, the division of property between heirs, combined with repeated sale and purchase, led to the continued splitting and re-splitting of common-field parcels until they could be split no more because they would have been too narrow to cultivate and often the holdings they formed part of too small to feed a family.[543] Therefore at some stage many peasants felt compelled either to abandon the partition of holdings between heirs, as they did in Syria, or to try to have no more than one heir, as they did in China and France.[544] Any custom of partible inheritance was thus not the cause of division between heirs, but merely the legal form given to such division, so that it could more truly be said that it was the division that gave rise to the custom. When division became undesirable, the custom of partible inheritance was eventually thrown over altogether or circumvented in some way. Only the rigid French legal code prevents peasants taking the obvious step of overriding custom by devising the holding by last will and testament. If all else fails and there are too many heirs for the holding, it has not been unusual for them to circumvent the law by coming to terms and, provided there is money enough, for one to buy the others out. Another option has been for the heirs to refrain from marriage, form a partnership and run the entire holding between them.[545]

The immediate palliative applied to excessive morcellisation has been consolidation. The proprietors exchange lands one with another and build up larger and more workable parcels.[546] Alternatively, some proprietors have succeeded in buying out others and consolidating all the parcels. Such *Flurbereinigung* (field-refining) and *Zusammenlegung* (laying together) have been encouraged by legislation in many German and other states, and recently have been financed by the European Economic Community.[547] But such consolidation in itself has provided no lasting remedy; it has often merely made it possible to recommence the division of holdings between heirs and purchasers, and so has helped to perpetuate common fields.[548]

Consolidation has often occasioned the rearrangement of furlongs to accommodate the larger parcels.[549] Much more thoroughgoing rearrangements of furlongs and fields, however, accompanied the extrication of demesne lands from the common fields, which occurred in northern France in the eighteenth century and in Prussia and Hungary in the nineteenth.[550]

When it comes to discussing the original causes of the creation of common fields, few Europeans give any credence to theories based on supposed co-aration. For one thing, no evidence of co-aration has been forthcoming.[551] Besides, where the plough was drawn by a single light horse or a donkey, a cow, a camel or a buffalo, and there was consequently no composite team to be made up by co-aration or any other means, common fields developed all the same.[552]

The origins of common fields are nowadays explained along some such lines as these. The starting point might be either an isolated homestead (*Einzelhof*) set in the wilderness and with a *Blockflur* consisting of mere patches of cultivation, or a hamlet with a tract of land divided lengthways between the peasants, so forming a *Waldhufenflur*, *Marschhufenflur* or other *Langstreifenflur* – a woodlanders', marshlanders' or other 'long-strip' field.[553] Such plots and patches, or the parcels within them, would either be in turf husbandry, as in Lower Saxony, or, more usually, would be tilled temporarily, the plough being shifted from exhausted soil to new plots in some kind of rotation, and each plot, which was in the nature of a furlong, being temporarily under some kind of field-course that perhaps incorporated bare fallows lasting a year or so.[554] The plots subjected to temporary tillage would be dispersed here and there, for the first settlers, like the English ones in south and east Africa in a later age, could pick and choose with all the fastidiousness of epicures the tracts of land thought most suitable for cultivation, and these were likely to be scattered in various places conveniently near to the homestead.[555]

Clearing a wilderness for this kind of cultivation (or, for that matter, building and maintaining seawalls) demanded great expenditures of labour, and life at the frontier was often dangerous, so safety might be sought in numbers. Partly for these reasons, each homestead or hamlet would come to be occupied by an extended or patriarchal family, called a *Hauskommunion* in German, a *manse, condoma* or *condomina* in Latin languages, a *zadruga* or some such in Slavic ones, and a *fellah* in Arabic. These families embraced three or four generations, and up to fifty, sixty or even a hundred persons, all living in the same place and working the same land, and all under the effectual rule of a single patriarch. The sons, even if fifty or more years of age, had no property of their own, for all was held in common, in a sort of primitive communism. The patriarch selected the tillage plots, oversaw the work, and divided the proceeds amongst his people as he saw fit.[556] Eventually, by dint of years and years of hard labour, parts of the land became sufficiently ameliorated, and the livestock numerous enough, to allow the introduction of some permanent tillage with bare fallows every second or even third year. All that could be adequately dunged or folded became permanent tillage, while temporary tillage plots tended to be banished to the more remote quarters. Since they developed from former temporary tillage plots, the areas of permanent tillage would, from the outset, have been in furlongs that bore traces of the initial scattering and so tended to be grouped in fields extending in this and that direction from the inhabited point. But permanent tillage in a system of mixed agriculture demands not the massed labour of an extended or patriarchal family, but management in smaller holdings by smaller and more cohesive groups, by conjugal families devoting loving care and attention to their own lands and beasts. As conjugal families formed and became increasingly attached to their own lands and livestock, and their menfolk resentful of patriarchs, so the patriarchal family dissolved into its constituent conjugal parts.[557]

With the decay of the patriarchal family, the hamlet became based less on consanguinity and more on territorial neighbourhood, and finally became a township, with a town meeting or some form of town government, to which the erstwhile patriarchs would be subordinated and from whose counsels they might even be excluded.[558] At the same time, the unit of land previously considered enough for one family – *terra unius familiae* – was divided up into smaller units, each of which, in permanent tillage, was enough for one conjugal family.[559]

European scholars give surprisingly little attention to the part possibly played by slavery in this whole development. Yet a supposed prior rise in population is often invoked as the cause of the grand division of landed possessions.[560] In fact, no one has so far been able to show any significant increase in population preceding the introduction of permanent cultivation that led to the splitting up of patriarchal lands. One would expect this introduction of a much more fruitful kind of husbandry to precede a rise in the number of people. It may well be that an increase in the population resulted from the innovation of permanent cultivation. We shall never know for certain. What we can say with confidence is that a prior increase in population cannot be shown to have caused the collapse of the patriarchal family and the splitting up of its lands in any particular place, and what cannot be shown in particular cannot be shown in general. Such slight evidence as we currently have points in the opposite direction. When Markišavci, in north-west Yugoslavia, was split up, some time between 1860 and 1936, the number of houses rose, but between 1869 and 1953 the population rose only from 202 to 206.[561] Anyway, the break-up of the patriarchal family is by itself ample reason for the splitting up of its lands.

The other main way estates were broken up was on a grander scale, according to a master plan, and in a regular and orderly fashion. Such division occurred when the slaves on an estate were hutted and transformed into service-tenants[562] or sharecroppers, each with his own allotment of land,[563] or when an association of colonists or an established township opened up new fields in virgin lands,[564] or when lords carved out new estates and set and let them to individual colonists.[565] In New England, where fields were laid out by colonists' associations, it was rare for any land to come into permanent cultivation, but this was because any young man could go west. The first agricultural frontier to roll across the continent was a moving frontier of temporary cultivation.[566]

In all these divisions of the lands of old or new settlements, each head of family was allotted what was considered his fair share of good, bad and indifferent land, of marsh and upland, of sandy and clayey soils, of shady and sunny and near and distant parcels. This meant in practice that where the fields and furlongs were already established, each furlong in turn was split between all the allottees. Only in those rare instances where the soil and the terrain were both more or less uniform was it possible to achieve an equitable division by any other means. As a general rule, then, when the lands were divided, each man was allotted his land in several parcels that were somewhat dispersed, scattered and intermingled with those of other men.[567]

The chief difference between the results of these two main lines of development was that the break-up of the patriarchal family and the division of its lands tended to lead to irregular common fields with less regular dispersal of the parcels, and the planned settlements to more regular ones with more regular dispersal. But always the initial lay-out of the original fields was decisive in determining the basic field pattern followed in later ages. What exactly resulted from division was determined by what exactly was to be divided. It was the natural environment that determined whether the fields were to become large or remain small and whether the furlongs were to run this way or that.[568]

And we should not assume that, at the time of the first great division, conversion to permanent cultivation had gone as far as it eventually did or that the introduction of permanent tillage led immediately to such a division. In most instances, permanent cultivation is more likely to have been introduced gradually, furlong by furlong. As long as the tillage was purely temporary and shifted from place to place, the location of parcels within it was necessarily temporary and the land was redistributed from time to time as occasion demanded. The introduction of permanent cultivation led, sooner or later, to the abandonment of redistribution. In India this change was apparently made with relative ease.[569] The interior of eastern and south-eastern Syria still has the *mouchaa* system, in which the temporary tillage is redistributed among the *fellaheen* (members of the patriarchal family). Formerly, these redistributions were made annually, but nowadays less frequently in many places, presumably because tillage is now prolonged into a second or third year. In the better favoured parts, first in Latakia, and then in the Bekaa and Aleppo plains, the *mouchaa stabilisé* has been introduced. The land having been brought into permanent tillage, it has been permanently allotted in common fields to the various conjugal families and is treated as their own property. How the *mouchaa* became thus stabilised is easily explained: those cultivators who had improved the parcels allotted to them refused to part with them in exchange for unimproved ones. Individual properties had been created by the investment of labour, effort and fertiliser, and the improvers saw to it that they enjoyed the fruits of their own labours.[570] Systems similar to the *mouchaa* are found in Turkey, where some are perhaps being stabilised, and still, perhaps, in Ethiopia.[571]

Likewise, in Russia and other parts of the Tsarist empire, temporary tillage periodically repartitioned was associated with the patriarchal family, and permanent tillage and permanent individual property with the conjugal family. In parts of north and Little Russia in the seventeenth century, the advent of permanent tillage in the possession of conjugal families was accompanied by the abandonment of redistribution. Elsewhere in the empire, though less strongly in the west and south-west, periodical partial or general repartitions continued after 1861, and all sales of repartitional land had to be approved by the *mir* or township. Only after 1905 did the area of repartitional land substantially decline, and only under a law of 1906, becoming effective in 1910, was repartition and temporary occupation abolished in favour of hereditary tenure. The Tsarist government put an end to repartioning because it was already being obstinately resisted by those who had improved their allotments and made them fit for permanent tillage. As might have been predicted, those who had failed to improve their portions were strongly in favour of repartition, arguing that the land was a gift of God and should be shared equally by all, which is perhaps true for nomads, but conveniently forgets that every man has a natural right to the fruit of his own labour and that while the wilderness is a gift of God, improved land incorporating capital is the work of Man. It was this leveller campaign, organised by townships dominated by the unthrifty and envious, that turned the Tsar's minister Stolypin against the *mir* as an institution.[572]

One thing all divisions and subdivisions had in common was the method of division itself. Innocent of advanced surveying, those who laid out new allotments or divided up old furlongs simply took yardsticks, rods, ropes or chains, measured along the ends or shorter sides, and split the land into long parcels. When furlongs were divided over and over again in this fashion, the parcels became ever narrower and their ownership ever more widely scattered and dispersed. In German parlance, *breitstreifige Gewannfluren* became *schmalstreifige*.[573]

In trying to discover when and where the permanent common fields of north-western Europe came into being, we have five indistinct trails to follow: field-courses, permanent cultivation, field systems, conjugal families, and service-tenancy. Only when two or more of these converge have we any hope of coming upon what we seek.

Some earlier German scholars, who seem to have influenced Gray, regarded the *Dreifelderwirtschaft* practised in *Gewannfluren* as a unique invention that sprang ready-made from the heads of Germanic people in the Dark Ages and was then imposed by them in the territories they won from the inferior, backward and primitive Slavic and Celtic races. This theory, understandably, appealed to many Germans.[574] Since they held that the *Dreifelderwirtschaft*, by which they meant the two- and three-field system in permanent cultivation, was exclusively Germanic, these theorists saw no obstacle to divining the presence of Celts or Slavs from the presence of temporary, and the absence of permanent, cultivation, in early times, or in turn to divining this presence or absence from nineteenth-century maps and plans. Meitzen had little hesitation in categorising the Kampen district in the east of the Netherlands as Celtic, which has since been shown to be pure fantasy.[575]

Worthy of closer consideration is the work of historians and others on the three-field system. Claims made for the existence of three-field courses in Carolingian Germany almost all relate to the upper Rhine valley, Moselland or thereabouts, and are thus irrelevant to our enquiry. Some of these courses may have had one year of crop followed by two of fallow, but all were assuredly of the peculiarly Roman kinds, unlike those known in north-western Europe.[576] Anyway, these claims are not always supported by the evidence. The earliest one relates to the year 763. A man called Hug leaves to the monastery of St Gall various lands in Weigheim, with

> *censum annis singulis, hoc est cervice siclas xx, maldra panis et frisginga saiga valente, et opera in stathum tempus, in messe et fenum duos dies ad messem medendum et foenum secandum, et in primum vir arata jurnalem unam, et in mense junio brachare alterum, et in autumno ipsum arare et seminare, hoc est censum pro ipsa villa.*

Hug grants the annual rent, that is, ale, bread and a sound young boar, and works at fixed times: two days' cutting and making hay, one day's ploughing in early winter, another stirring the land in June, and ploughing and sowing one day-work in autumn, this being the standard rent in the villa. This shows that winter corn was grown on land previously broken up and stirred; but that ploughing was done in three seasons proves nothing as to what field-course was used. Weigheim and other towns on this estate are outside our main field of interest; and we note in passing the typical southern practice of pulverising the soil in summer. We are also invited to consider as good evidence of the three-field course a passage relating to Thuringia in 783 and reading *in tribus Hoheimis ... in tribus Geochusis ... in tribus Percubis*. This is really a gobbet from a grant to the Benedictines by Abbess Emhilt of her lands in Milz, in the three Höchheims, in Sülzdorf, in the three Jüchsens, in the three Berkachs, in Wielantesheim, Hellingen and other towns, together with all their movable and immovable goods. It so happens that these lands were regranted in 796, and this time the editor gives the town names rather better, 'Geochusis' now being rendered as 'Juchisis' (Jüchsens) and 'Percuhis' (which, incidentally, White changed to 'Percubis') as 'Berchosis' (Berkachs). Now, too, the editor transcribes the names of all the towns, which include, as well as the three Berkachs, three Jüchsens and three Höchheims, the two Essfelds, Behrungen, Römhild, Hindfeld, Döringstadt, Themar and over a dozen other towns and hamlets flanking the Thüringerwald or Thuringian Forest, and mostly lying just south-west of it. The documents speak of the two Essfelds, the three Berkachs and so on, just as we would of the two Ockendons or the three Woodfords. Neither fields nor field-courses are anywhere mentioned.[577]

And when we are told that by ' "130 Joch in der Villa" ... kann nur die Dorfflur gemeint sein', we must take *Dorfflur* in its widest and loosest sense and not to mean a town or common field where the demesne lands were intermixed with those of the tenantry.[578]

Not to be outdone, French historians have also advanced extravagant claims. That on one part of the Parisian abbey of St Germain-des-Prés about AD 812 a tenant *Arat ad hibernaticum perticas iiii, ad tramisum perticas ii* – 'ploughs for winter-seed 4 perches, for spring-seed, 2' – proves neither permanent cultivation nor any field-course, far less a three-field one, especially as twice as much was ploughed in winter as in spring.[579] This same phenomenon, of twice as much ploughing in winter as in spring, is found almost throughout the abbey's estates in the Ile-de-France. In several places plough-works were demanded in perches, ridges or day-works, without any mention of seasons.[580] In le Corbonnais most occupiers sowed to halves, as sharecroppers.[581] Otherwise, apart from idiosyncrasies such as found on all estates,[582] most of the tenants were bound to plough twice as much for winter as for spring crops, commonly at the rate of four to two.[583] In several places occasionally, but more generally in Combs-la-Ville, Chavannes and St Germain-de-Secqueval, three perches were ploughed for winter and two for spring crops or four for winter and three for spring, or at such rates.[584] A few tenants here and there ploughed as much for spring as for winter,[585] and even a few more for spring than for winter, in one or two instances much more.[586] But those who ploughed only for winter-seed and not at all for spring were far more numerous.[587] Only a small part of the arable seems to have been tilled at any one time. Demesne arable was measured in bonniers and tenants' plough-works usually in mere perches. In Celles-les-Bordes, for example, the demesne arable was sixty-five bonniers, of which only a small fraction was worked by service-tenants.[588] Anyway, ploughing twice as much for winter as for spring-seed is no proof, but rather disproof, of a three-field course. The sole indication of field-courses comes from Boissy-en-Drouais. Here the standard plough-work was four perches equally at winter-seed, at spring-seed, and *ad proscendendum* – 'at breaking up land' – implying land newly brought under the plough. This suggests temporary, shifting cultivation, as does the growing of a very large proportion of spelt, which is an inferior grain sown in poor land.[589]

Nor does *herbaticum* being paid every third year prove anything to the point. The tenants paid not for *herbaticum* every third year, but every third year for *herbaticum*, just as they paid *carnaticum* every third year and gave porkers every third year and shingles every fifth. The relevant entries in the polyptych are all much like this: *Ad tertium annum, propter herbaticum, germia I* – 'At the third year, for herbage-due, one theave' (an ewe ready to yean for the first time). This suggests in-lamb ewes were put into the meadows, in accordance with what has always been good standard practice. *Herbaticum* is a payment for either mowing or depasturing grass. We are not told, but if the *herbaticum* was for mowing grass for hay, that would be irrelevant here. Guérard, the worthy editor, would have preferred to conform to the belief prevailing among his contemporaries that the three-field system came from time immemorial, and tries to argue that *herbaticum* included *pascuarium* and was due for depasturing on the fallow field, but his scholarly conscience compelled him to confess that *pascuarium* was strictly quite distinct from *herbaticum*, which therefore 'à ce qu'il semble, ne s'entendre que du droit de faire de l'herbe'. Despite all this, merely for argument's sake, let us assume *herbaticum* was taken on the shack or fallow in a three-field course. If on the shack, then this recurs on two-thirds of the land in every single year, not just in one year in three; if on the fallow, then, this recurs not every third year but every single year. Thus, however *herbaticum* be taken, this passage is no proof of any kind of field-course.[590]

It is claimed that the horse-collar, the 'deep' plough, and the three-field system were all introduced in the Paris basin in the late eighth and early ninth centuries; but we are told neither if the plough had a mould-board, nor the exact nature of the field-course, nor whether this course was in permanent or temporary cultivation. We could not even accept an assertion that three-field courses were employed in temporary cultivation. When we are told that the distinction between spring and autumn sowing betokens an embryo three-field course, we can only gasp in amazement. All the distinction shows is that spring and autumn crops were distinguished from one another. The few full, clear and authentic records of any field-course in this region in the twelfth century give evidence not of permanent but of temporary and shifting cultivation.[591] Incidentally, we need to be wary of the word *jachère*, for it can mean either that the land was fallowed or that it was abandoned to the wild for up to ten years or so.[592]

Professor Duby once claimed that the three-field course was found in the ninth century in what he was pleased to call the heartland of medieval civilisation, namely, the plain of France between the Loire and the Rhine, but showed no evidence and gave no clue to the exact nature of the course or to whether it was in permanent or temporary cultivation. As for his assertion that the three-field course made it possible for the first time to grow oats for horses, why, oats were much grown in temporary, shifting tillage plots, with or without any field-course. Oat cultivation depends not on any particular field-course but on rain in spring and early summer.[593] Later he confessed that only one document from the ninth and tenth centuries showed a three-field course of (1) winter corn, (2) spring corn, (3) fallow.[594] But this fragment, unfortunately, refers to the estates not of the abbey of St Germain-des-Prés, but of the abbey of St Amand, which was not even in France at all, but in the Netherlands, between Doornijk and Valenciennes.[595]

M. Duby says, 'On a la certitude qu'un rhythme semblable était généralement suivi en Normandie dans la deuxième moitié du XIIIe siècle', and cites in sole support a learned article by Professor Strayer.[596] But this makes no mention of anything in any way connected with field-courses. The nearest thing to such a mention is contained in the particulars of a grant in three parts, 'la tierce partie en grains, la tierce partie en domaines et en rentes, et l'autre tierce partie en homages et en seignuries.'[597] Elsewhere, however, Professor Strayer has made an allusion to a three-field course. In his work on the *bailliage* of Rouen, he writes,

> It seems probable that the three-field system was generally used in the *bailliage*. Horses, rather than oxen, were used for plowing, a practice which usually indicates the three-field system. In addition, there are definite references to winter and spring sowings in the viscounty of Pont de l'Arche. The typical list of *corvées* states that plowing is due twice a year and harrowing twice a year.[598]

This is an unfortunate momentary aberration. Horse-ploughing is not indicative of any field-course. Winter and spring sowings show only that the location is more likely north-western than Mediterranean Europe. Ploughing and harrowing twice a year is, of course, common practice in both. Such is the stuff of M. Duby's 'certitude'.

The very first evidence so far found of a three-field course in permanent cultivation on the Continent is found in the fragment from the early ninth-century polyptych of St Amand abbey already referred to. In the villa Businiacas (present-day Bousignies), in Hainaut, we read, *Ad mansum dominicatum pertinant de terra arabili bunaria xi. De his seminantur ad hibernaticum bunaria v, de modiis xx; ad aestivum, bunaria vi, de modiis xxxvi.* – 'To the demesne messuage belong, of land arable, 11 bonniers. Of these are sown at the winter-seed 5 bonniers, with 20 pecks; in summer, 6 bonniers, with 36 pecks.' Immediately after '*bunaria xi*' is an editorial note '[*potius xvi*]' – '[16 rather]'; but this is gratuitous, for 5 and 6 make 11.[599] Guérard edited the text to make it fit his belief, pardonable in 1844, that three-field courses were normal in early times. Five bonniers were sown in winter and six in summer. Spring sowings and twelve-month fallows are not mentioned. Apparently, winter and summer cropping was not alternated between the five bonniers and the six; the one part was always sown in winter and the other in summer. Possibly this land was in a three-field course, but more likely it was every year's land, subjected to an endless succession of crops.

In the five bonniers, fallows could scarcely have been made, but in the six, winter fallows would have been made every year. Assuming the heavy manuring usually found in industrial districts like this, large acreages of podware, and the rotation of different crops in each of the winter and summer fields, this kind of every year's land appears feasible.

The next entry is quite different. It concerns the villa Madria, where there is a *mansus dominicatus cum casa et ceteris aedificiis, cum orto, habens de terra arabili bunaria xxx. Seminantur ad hibernaticum bunaria x, de modiis xl, et ad tremissem bunaria x, de modiis lx. Bunaria x interjacent.* – 'demesne messuage with a cottage and other buildings, with a garden, having of land arable 30 bonniers. There are sown at winter-seed 10 bonniers, with 40 pecks, and at spring-seed, 10 bonniers, with 60 pecks. 10 bonniers lie between.' But probably *inter* and *jacent* should be separate words, and then the meaning would be, more reasonably, '10 bonniers meanwhile lie fallow'.[600] Here we have clear evidence of a three-field course, apparently of the downland type.

The next entry relates to an unknown villa. It reads, *Ad mansum dominicatum pertinant de terra arabili bunaria xlviii. Ex his seminantur ad hibernaticum bunaria xvi, de modiis lxiiii, et ad tremissem bunaria xvi, de modiis lxxxxvi.* – 'To the demesne messuge belong of land arable 48 bonniers. Out of these are sown at winter-seed 16 bonniers, with 64 pecks, and at spring-seed, 16 bonniers, with 96 pecks.' The other third was obviously being fallowed. Here again we have a three-field course.

Lastly, at Salaconis, the villa Millio has *de terra arabili bunaria xv. De his seminantur de sigali ad hibernaticum bunaria v, de modiis xx; ad aestivum bunaria v, de modiis xxx.* – 'of land arable 15 bonniers. Of these 5 bonniers are sown in winter with rye, with 20 pecks; in summer 5 bonniers, with 30 pecks.' The other five bonniers evidently lie fallow. Once again we have a three-field course, though, it seems, with summer instead of spring sowings.[601]

Although no evidence has been found of field-courses with fallow years before the thirteenth century in Flanders, we suppose it at least kept pace with other parts of the Low Countries.[602]

Possibly three-field courses were followed in parts of Picardy somewhat earlier,[603] but the first sure evidence, for the Paris basin, of a field-course incorporating fallow years in permanent tillage, comes from 1248, when it is recorded,

Sciendum quod totum territorium Vallis Laurencii dividitur in tres aristas. Prima arista segetis continet xviii^{xx} *et v arpennos et dimidium et vi perches. Secunda arista que est in iaquera continet xvi*^{xx} *et iii arpennos et ix perches. Tercia arista que est in marcesche continet xvi*^{xx} *et xiii arpennos et x perches.* – 'It should be known that the whole territory of Vaulerent is divided into three seasons. The first season contains 365½ arpents and 6 perches. the second, which is in fallow, contains 323 arpents and 9 perches. The third, which is in spring crops, contains 333 arpents and 10 perches.'

This was on permanent tillage in severalty at Vaulerent, a Cistercian grange near Louvres and just east of the Senlis to Paris road. Possibly, this course was already being followed here in 1194.[604] In la Beauce in 1225 and at Marly-la-Ville in 1237, three-field courses were operated, some perhaps in permanent tillage.[605] In 1252 some villages in Avesnes canton in Artois followed similar courses, though whether entirely methodically and in permanent cultivation is not absolutely clear.[606] Just possibly a two-field course was used at Saint Saëns in Normandy in 1257.[607] A three-field one was being worked at Dabeuf-la-Campagne, in the plain of Eure, in 1291.[608] By this time three-field courses seem to have become general in Artois, even though not always in permanent tillage.[609] By the fourteenth century two- and three-field courses had been adopted in various parts of the Norman plain and the Paris basin; but, before the sixteenth century, evidence of these courses in this region is extremely rare and not always wholly convincing; and nowhere, unless perhaps in Artois, does it suggest that these courses were being followed in common fields.[610] Moreover, it seems that two- and three-field courses, and permanent cultivation itself, rarely survived the Black Death and other calamities in the fourteenth century and devastations in the fifteenth.[611] The general introduction of two- and three-field courses in permanent common fields came only in the early modern period, and then notably in the Ile-de-France.[612]

The earliest sure evidence for field-courses incorporating fallow years in permanent tillage in Germany comes from the vicinity of Cologne in 1251 and 1277. Here, and more generally in the vale of the lower Rhine, at Düren and elsewhere, such courses were found in the fourteenth century. In every instance, the evidence relates to the growing of horsemeat or podware specifically in the fallows.[613] Hanssen draws our attention to the lessee at Sulz who in 1250 was obliged to sow vetches *in agris qui illo anno non erunt seminandi* – 'in the fields not due to be sown that year'. Hanssen cautiously refrains from concluding that a three-field course or *Dreifelderwirtschaft* was being practised here,[614] perhaps because he thought fallows were not referred to clearly enough. He may also have recalled how poorly endowed for agriculture are the Sauerland and Siegenland. But Sulz, if the town of that name in Kreis Siegen, being over by the Rhine, would likely have been exceptionally fertile. On balance, the passage probably refers to a two or three-field course in permanent tillage.[615] Yet, although instances of seasons with fallows, some perhaps in permanent tillage, are found earlier, it appears it was not until the sixteenth, seventeenth and eighteenth centuries that two- and three-field courses in permanent tillage spread generally in northern Germany as a whole. It was an early modern development,[616] as it was also in Denmark and in suitable places in the rest of Scandinavia.[617] Poland followed on the heels of Germany,[618] and so did Latvia, Estonia and Lithuania.[619] What three-field courses obtained in Russia up to the nineteenth century were mostly in temporary, shifting cultivation, and permanent tillage with field-courses incorporating fallows was generally achieved only in that century.[620] Thus, roughly speaking, this piece of husbandry spread step by step from west to east across the great European plain.[621]

The point has been well made, that the rise of the German *Gewannflur*, or field composed of furlongs, more or less coincided with the rise of *Dreifelderwirtschaft*, or two- or three-field system.[622] But only where permanent tillage coincides with fields composed of scattered parcels belonging to various individual cultivators can we speak of permanent common fields. We need to look for such dealt or parcelled-out fields. The present German consensus is that such *streifige Gewannfluren*, or furlonged 'strip' fields, and particularly *schmalstreifige Gewannfluren*, with narrow 'strips', though they may have had their beginnings somewhat earlier, only started to spread widely in the early modern period, approximately from 1560 to 1760, and only attained their greatest extent and mature form in the late eighteenth and early nineteenth centuries, in step with great rises in the numbers of cultivators working part of their time in manufacturing, processing, and extractive industries and associated trades, in the course of the wider diffusion of the so-called 'Industrial Revolution'.[623] Developments in more easterly parts of the great plain, and in Russia generally, seem to have taken a similar turn, with furlonged fields and regular field-courses arising hand in hand.[624] The same seems to be true of Yugoslavia after 1850.[625] Although the evidence is of a different kind and very meagre, it appears that permanent common fields finally took hold in most of northern France only in the seventeenth century and failed to expand to any great extent until the eighteenth, when large sheep-walks were laid out and the systematic folding of sheep on the tillage was started for the first time.[626]

Some confirmation of this chronological sketch can be dimly discerned in the break-up of patriarchal families and their holdings and the formation of service-tenancies by the hutting of the slaves and the enserfment of free peasants, all these events being roughly coeval. In Drenthe and the eastern Netherlands, they date from about the thirteenth century. The patriarchal family survived in Norway until the thirteenth century, in Sweden until the fifteenth, in Denmark and north and central Germany up to and beyond the sixteenth.[627] In northern France, the division of patriarchal families and holdings occurred in later medieval times, whereas in the main body of Germany it started only at the end of the Middle Ages, attained little momentum before the early modern period, and dragged on into the early nineteenth century.[628] The dissolution of the patriarchal family in Russia started in the middle of the eighteenth century, but only attained great momentum after 1861. In Yugoslavia it started in the middle of the nineteenth century; and in Bulgaria somewhat later. Even today the patriarchal family survives in much of southern Europe.[629] The hutting of slaves took place in Lithuania in the sixteenth century. It started in Russia in the later eighteenth century, but became general only in the nineteenth. In Romania and Bulgaria, hutting commenced earlier than in Russia, but was still going on in the nineteenth century.[630]

If we limit them to the northerly region we are chiefly concerned with, we may readily accept the conclusions of modern German scholars, that there was no *Dreizelgenbrachwirtschaft*, no three-field course with a fallow year, in Germany in Carolingian times,[631] and that 'Es gibt kein urgermanische Gewannflur', there is no ancient, original German furlonged or common field.[632] The ancient Germans knew only shifting, temporary cultivation. The words of Tacitus, *Arva per annos mutant, et superest ager* – 'The tillage plots are changed from year to year, and there is land and to spare' – can only mean that the Germans shifted their tillage from place to place in the wilderness, and fit in with his remarks that they troubled neither to fertilise the land, nor to work it thoroughly, nor to mark off any of it for meadow, but put successive fields into tillage. And, turf husbandry excepted, the northern Germans still mostly cultivated in the same way throughout the Middle Ages.[633]

But our knowledge of this serves to isolate and highlight the true and great achievements of the early Germanic peoples. They acquired one great and unique asset: they came to make, own and handle sulls, i.e. ploughs with iron coulters and shares and with wooden mould-boards that could turn a furrow-slice in the heaviest and wettest of soils and till them in true ridge-and-furrow fashion.[634] Both shifting cultivation and turf husbandry demand huge areas of land per person, so it was only a matter of time before the Saxons and other Germans experienced the land-hunger that led them in the twelfth century to embark on their *Drang nach Osten* and carve out great estates for themselves in the Slavic lands they conquered. Thanks to their sulls and their ridge-and-furrow techniques, the Germans were able to bring into temporary cultivation the wet and low-lying lands that the Slavs and their primitive wooden scratch-ploughs were unable to till properly, while the Netherlanders who took up the refrain 'Naer Oostland willen wy ryden' were well able to master the coastal and riparian marshes. Hence the contrast between the irregular, small *Blockfluren* of the Wends and other Slavs, nothing better than rudely cultivated patches dotted about the wilderness, and the larger, regular ones laid out by the Germans. It was from these German colonists that the native Slavs learned to plough ridge-and-furrow and turn over their soils. Then they followed the example set by the Germans and joined them in the taming of the wilderness.[635] By the end of the middle ages the Germanic peoples had thus prepared the ground and cleared the way for permanent cultivation and common fields throughout the great European plain.

Conclusion

What little we have learned about common fields overseas tends to confirm our conclusions about the English ones. Some points, indeed, make our surmises less tentative than they were. It now seems more credible that temporary cultivations were converted to permanent ones by the progressive diminution of the area abandoned to the wilderness after a succession of two- or three-field courses.

Many parallels may be drawn between England and Europe, but perhaps the most interesting is that between the achievements of the Germanic peoples in the Celtic and Slavic lands they took and occupied. They took their ploughs and their ploughing techniques with them wherever they went, and cleared the wildernesses that more primitive peoples were by themselves unable to bring into cultivation. Yet for a long time these Germanic pioneers engaged in no more than shifting and temporary tillage, which none the less prepared the way for permanent cultivation.

From all the evidence that European scholars have so far managed to unearth, it appears that permanently cultivated common fields may have started in the Low Countries and the Lower Rhineland, and even in part of the Paris basin, somewhere about the thirteenth century, but elsewhere in the northern French and German plains the rise of such common fields was an early modern phenomenon. The further east we go, the tardier the development of permanently cultivated common fields. Their introduction formed a moving frontier that proceeded roughly and irregularly from west to east. And this late development of common fields in Europe helps to explain their late continuance.

England, as usual, was different. She had common fields (and, presumably, the sull), in the eighth century, and permanently cultivated common fields in the tenth.[636] Clearly, the English were as precocious in their agriculture as in their centralised government. One possible reason for this springs to mind. By and large England's soils and climate were more suited to cereal cultivation than were

Germany's. But the most important technical reason is that the English invented and mastered the arts of sheep-and-corn husbandry. The crucial role of the sheepfold in the creation, extension and maintenance of permanent tillage is not always fully appreciated. It was only when the northern French learned from the English the art of folding sheep on the arable that they were able to create a large extent of permanent tillage.[637] The special significance of the common fields in the early system of permanent cultivation is that without them and their tenants, the demesne lands could hardly have been kept in permanent tillage either, so the whole system depended on common fields.

England's precocity in agriculture had consequences of much wider significance. Throughout all the Middle Ages, her system of permanent cultivation, in and out of common fields, was based on the folding of arable sheep. At a time when Flemish sheep gave only a soft and curly hair and French ones something akin to dog's hair, England had developed breeds of arable sheep with fleeces of good fallow wool, and this first her clothiers, and later, and until the fifteenth century even more successfully, the Flemish ones, made into a wholly new type of fine cloth. Fallow wool, processed or unprocessed, gave England her staple export; it was the one result of advanced English agricultural technique that foreigners were able to experience at first hand.[638] From the ninth century or somewhat before, fine English cloth, later English wool, and, from the fifteenth century, English cloth again, along with some wheat and base and semi-precious metals, earned a lion's share of the silver mined in south Germany, and this, together with her own far more modest output, gave England silver and to spare. She, almost alone, generally enjoyed honest silver money.[639] Thanks chiefly to her agricultural innovations, England became wealthy out of all proportion to her size and population. This made her a great temptation to aggressors, but enabled her kings to maintain a strongly centralised government, to contain and largely subdue the tribesmen on her western and northern borders, and to subsidise allies in Continental wars.[640]

The English invention of a system of permanent common fields suited to the more northerly parts of Europe ranks as one of the major events in the history of the western world.

References

The place of publication is London unless otherwise stated.

1 W. Marshall, *Rev. (and Compl. Abs.) Reps to B. Ag. fr. S. and Pen. Depts Engld*, 1817, pp. 166, 264; C.R. Straton, *Surv. Lds Wm, 1st E. Pemb.*, 2 vols, Rox. Club, 1909, p. 242; E. Kerridge, *Survs Mans Phil. E. Pemb. and Montgom. 1631–2* Wilts. Recs Soc. ix, 1953, p. 2.

2 E. Kerridge, *Ag. Rev.*, 1967, pp. 16–19; R.H. Hilton, *Med. Soc.: W. Midlds at end 13th cent.*, 1967, p. 19.

3 W. Fream, *Elems Ag.*, 1918, pp. 54–7; R.N. Salaman, *Hist. and Soc. Infl. Potato*, Cambridge, 1949, pp. 46, 48; T. Hennell, *Change in Fm*, Cambridge, 1934, pp. 59 sqq.; R. Plot, *Nat. Hist. Oxon.*, Oxford, 1677, pp. 239–41; Fitzherbert, *Boke of Surveyinge and Improvementes*, 1535, fos 43, 44v.

4 Ibid., fo. 43; my *Fmrs O. Engld*, 1973, pp. 29–32; 'Ridge and Furrow and Agrarian Hist.', *Econ. Hist. Rev.* 2nd ser. 1951 iv, 15.

5 Ibid., 15, 16; my *Fmrs*, pl. 1; Bodl. Hearne's Diaries 159 p. 256; W. Marshall, *Expts and Obs. con. Ag. and Wthr*, 1779, intro.; *Rev. Reps to B. Ag. fr. E. Dept Engld*, 1811, pp. 435–6; *Rural Econ. Nfk*, 2 vols, 1787, i, 131.

6 Ibid., 147; *Rural Econ. S. Cos*, 2 vols, 1798, ii, 322; I.B. in Fitzherbert, *Bk Husb.* ed. W.W. Skeat, E.D.S., 1882, p. 132 (quot.); J. Morton, *Nat. Hist. Northants.*, 1712, p. 55; R. Plot, *Nat. Hist. Staffs.*, Oxford, 1686, p. 340; E. Lisle, *Obs. in Husb.*, 1757, p. 71; J. Boys, *Gen. View Ag. Co. Kent*, 1796, p. 56; W. Blith, *Engl. Improver Improved*, 1652, p. 103; my 'Ridge and Furrow', 16–18.

7 Ibid., 18, 19; W. Folkingham, *Feudigraphia: synopsis or epitome of surveying*, 1610, p. 48; W. Cobbett, *Rural Rides*, 2 vols, E.E. ii, 41.

8 My 'Ridge and Furrow', 19–22; P. Kalm, *Kalm's Acct his Visit to Engld*, 1892, pp. 165, 271–2, 279; E. Maxey, *New Inst[r]uction of Plowing and Setting of Corn*, 1601; W. James and J. Malcolm, *Gen. View Ag. Co. Buckingham*, 1794, p. 19; J. Worlidge, *Systema Agriculturae*, 1669, p. 32; Blith, op. cit., 10, 103, 105, 117; J. Mortimer, *Whole Art Husb.*, 1707, p. 48; W. Marshall, *Rural Econ. Midl. Cos*, 2 vols, 1790, i, 221; *S. Cos*, i, 350; ii, 142, 235; *E. Dept*, 135; *Mins, Expts, Obs. and Gen. Rems on Ag. in S. Cos*, 2 vols, 1799, i, 82–3, 118, 180; W. Ellis, *Mod. Husbn*, 8 vols, 1740, i, 48; ii, 49, 50; *Chiltern and Vale Fmg Explained*, [1733], pp. 23, 307; Plot, *Staffs.*, 340; *Oxon.*, 239–41; J. Laurence, *New Syst. Ag.*, 1726, p. 62; G. Markham, *Engl. Husbn*, 1635, pp. 37–8, 42; *Inrichment Weald Kent*, 1625, p. 16; Lisle, op. cit., pp. 53, 64, 121, 325; Fitzherbert, *Boke of Husbandrie*, 1523, fo. 8.

9 Bodl. Hearne's Diaries 159, p. 256; W. Marshall, *Rural Econ. Glocs.*, 2 vols, Gloucester, 1789, i, 56, 75–7, 83–4; *Rural Econ. W. Engd*, 2 vols, 1796, i, 32; ii, 208; J. Norden, *Surveiors Dialogue*, 1618, p. 228; W. Palin, 'Fmg Ches.', *Jnl R. A. S. E.* 1844 v, 62–3, 77–8; J. Bravender, 'Fmg Gloucs.', ibid. 1850 xi, 151; C. Vancouver, *Gen. View Ag. Co. Devon*, 1808, pp. 140, 151, 155, 157, 167; J. Bailey, *Gen. View Ag. Co. Durh.*, 1810, pp. 103–4; J. Farey, *Gen. View Ag. Co. Derby*, 3 vols, 1817, ii, 96; G. Kay, *Gen. View Ag. N. W.*, Edinburgh, 1794, Flints., p. 8; W. Pomeroy, *Gen. View Ag. Co. Worc.*, 1794, p. 41; J. Billingsley, *Gen. View Ag. in Co. Soms.*, 1794, p. 158; 1798, p. 279; C. Hassall, *Gen. View Ag. Co. Carm.*, 1794, p. 15.

10 Blith, op. cit., 59, 61; H. E. Hallam, *Set. and Soc.: stud. early agrarian hist. S. Lincs.*, Cambridge, 1965, pp. 142, 145–7, 150–3.

11 Folkingham, op. cit., 48; Kalm, op. cit., 43, 151, 157, 165, 184, 204; Ellis, *Chiltern and Vale*, 310; Mortimer, op. cit., 48; J. Laurence, op. cit., 61; Worlidge, op. cit., 191; Plot, *Oxon.*, 240; Morton, op. cit., 401; Bailey, op. cit., 103–4; Marshall, *E. Dept*, 436; *Nfk*, i, 131, 147; ii, 391; A. Kingston, *Hist. Royston*, 1906, p. 113; Sfk RO (Bury), Ep. Commy Ct f. Bury St Edms, inv. 1701/49; Herts. RO, Pym Coll. 19271, Ct R. Norton, 1 Oct. 1621; PRO, S.P.D., Jas, vol. 93, no. 125.

12 My 'Ridge and Furrow', 23–4; Plot, *Oxon.*, 240; Folkingham, op. cit., 48; Marshall, *Midl. Dept*, 596.

13 Bodl. Hearne's Diaries 159, p. 256.

14 Marshall, *Midl. Cos*, ii, 46.

15 My 'Ridge and Furrow', 26–8; Blith, op. cit., 100; Pomeroy, op. cit., 41; W. J. Slack, 'Open Fld Syst. Ag.', *Trans. Caradoc and Severn Val. Fld Club* 1937 x, 153–4; J. Crutchley, *Gen. View Ag. Co. Rut.*, 1794, pp. 12, 13; James and Malcolm, *Bucks.*, 19; W. Gooch, *Gen. View Ag. Co. Cam.*, 1811, p. 139; St J. Priest, *Gen. View Ag. Bucks.*, 1813, p. 131; J. Middleton, *View Ag. Mx*, 1798, p. 139; W. Rider-Haggard, *Rural Engld*, 2 vols, 1906, ii, 69, 247.

16 C. Vancouver, *Gen. View Ag. Co. Cam.*, 1794, pp. 121, 148; Shak. Bpl Throckmorton Coll. Ct R. Oversley 13 Oct. 1762; my 'Ridge and Furrow', 33–4; cf. Marshall, *Midl. Dept*, 634–5.

17 My 'Ridge and Furrow', 25–7; Fitzherbert, *Surveyinge*, fo. 44; Blith, op. cit., 103–5; Marshall, *Mins*, i, 82–3; Farey, op. cit., ii, 96; Ess. RO, D/DP E25 fos 61v, 70; W. R. Emerson, 'Econ. Dev. Estes Petre Fam. in Ess. in 16th and 17th cents', ts. thes. D.Phil., Oxford Univ. 1951, p. 265; D. Woodward, *Fmg and Memo. Bks Hy Best Elmswell, 1642*, n.d., p. 34.

18 Marshall, *Midl. Cos*, i, 221; Markham, *Engl. Husbn*, 38–9; Worlidge, op. cit., 32–3.

19 My *Ag. Rev.*, 257–8; *Fmrs*, 112 and pls 18–22.

20 My 'Ridge and Furrow', 28, 32, 35–6; R. Brownlow, *Reports (A 2nd Part) of Diverse Famous Cases in Law* (2 Brownl. and Golds.), 1652 and var. eds, p. 46; Marshall, *Midl. Dept*, 17; *Glos.*, ii, 93; *W. Engld*, i, 32; ii, 45–6, 172; Morton, op. cit., 45; 'Proc. in St. Ch. Hy VIII and Ed. VI', *Colls Hist. Staffs.* 3rd ser. 1912, p. 184; W. R. Mead, 'Ridge and Furrow in Bucks.', *Geog. Jnl* 1954 cxx, 41–2; J. E. Stocks, *Market Harboro. Par. Recs 1531–1837*, 1926, p. 313; Rider-Haggard, op. cit., ii, 122–3; L. A. Harvey and D. St Leger-Gordon, *Dartmoor*, 1953, f. p. 7; P. Thompson, L. McKenna and J. Mackillop, *Ploughlands and Pastures: imprint agrarian hist. in 4 Ches. tps – Peckforton, Haughton, Bunbury, Huxley*, Chester, 1982, pp. 3sqq.; PRO, Exch. K.R., Deps by Cmmn,

21–2 Eliz., Mich. 28, dep. F. Ayleworth; 6 Chas, Mich. 12, ex parte quer., art. 11; 10 Chas, Trin. 4, m. 8; 13 Chas, Mich. 29, ex parte quer.; Mich. 59, arts 4 and 7; St. Ch. Proc. Jas, 55/29, mm. 2, 4, 8; Sp. Colls, Maps and Plans, MR/13; Chanc. Proc. ser. i, Jas 1, M. 7/4, complt; BL, Harl. MS 71 fo. 56 (67); Northants. RO, F(M) Coll. Misc. Vol. 434.

21 Leic. Mus. MS 35/29/272.

22 Marshall, *Glos.*, i, 75–7; my *Fmrs*, pl. 1.

23 'Proc. in St. Ch.', *Colls Hist. Staffs.*, 1910, p. 70; Kalm, op. cit., 158; Worlidge, op. cit., 212; H. Holland, *Gen. View Ag. Ches.*, 1808, p. 128; J. Laurence, op. cit., 63; Fitzherbert, *Surveyinge*, fo. 44; Northants. RO, Aynho Ct R. 20 Oct. 1685, 1 Apr. 1687; Shak. Bpl., Man. Docs, Ct Bk Alveston and Tiddington beg. 20 Apr. 2 Jas, 15 Apr. and 21 Oct. 9 Jas; Ct R. id. 11 Apr. 18 Chas 2; my 'Ridge and Furrow', 21; 'Reconsideration some fmr Husb. Practices', *Ag. Hist. Rev.* 1955 iii, 27–8.

24 Ibid., 28–9; W. Marshall, *Rural Econ. Yks.*, 2 vols. 1788, ii, 333; A. Stark, *Hist. and Ants Gainsborough*, 1843, p. 91; G. Eland, *Purefoy Ltrs 1735–53*, 2 vols, 1931, ii, 437; W. O. Ault, 'Open-Fld Husb. and Vil. Commun.: study agrarian by-laws in med. Engld', *Trans. Am. Phil. Soc.* n.s. 1965, lv, pt vii, p. 93; Mx RO, Acc. 446, M. 102, Ct R. Harmondsworth: 2 Oct. 1635; M. 104, 5 Feb. 1639; Shak. Bpl., Man. Docs, Ct Bk Alveston and Tiddington beg. 20 Apr. 2 Jas: ords 23 Oct. 7 Jas, 15 Apr. 9 Jas; Ct R. 11 Apr. 18 Chas 2; Berks. RO, D/EC/M. 151 Ct R. Hagbourne 2 Oct. 19 Chas 2; D/EPb M. 3 Ct Bk Coleshill 1592–1687, p. 66; Bucks. RO, Acq. 35/39 By-laws and Regs f. Padbury Com. Flds 1779; Cambs. RO, L.88/4 Ct files Caxton, presentments Whit 1659; Northants. RO, Mont. Coll., Northants. old box 7, no. 72, Ct Bk Weekley 16 Oct. 18 Eliz.; Misc. Led. 145, p. 184; PRO, Chanc. Proc. ser. i, Jas 1, S. 29/7 m. 1.

25 My 'Reconsideration', 29, 30.

26 Ibid. 29; Shak. Bpl., Throckmorton Coll. Ct R. Coughton 10 Nov. 1708; Leigh Coll. Ct R. Adlestrop 26 Apr. 1762; Herts. RO, Broxborne Bury Coll. 58, p. 112; S. P. Vivian, *Man. Etchingham cum Salehurst*, Suss. Rec. Soc. 1953, liii, 203; E. W. Crossley, 'Test. Docs Yks. Pecs' in *Miscellanea vol. ii*, Yks Archaeol. Soc. Rec. Ser. 1929, lxxiv, 100.

27 Blith, op. cit., 28 and f.p. 69; Worlidge, op. cit., 212; Soms. RO, Mildmay Coll., Acc. no. C. 186, box 6 Ct R. Q. Camel 20 Oct. 1701.

28 J. C. Halliwell, *Anc. Invs Furn., Picts, Tapestry, Pl. etc. illus. Dom. Manners Engl. in 16th and 17th cents*, 1854, p. 40; E. Laurence, *Duty of a Steward to his Lord*, 1727, p. 27.

29 My 'Reconsideration', 29, 30; 'Ridge and Furrow', 21; and as nn. 112–14.

30 Bateson et al., *Hist. Nthbld*, 15 vols, 1893–1940, iii, 101; PRO, Chanc. Proc. ser. i, Jas, Y.1/25; Exch., L.R., M.B. 210 fos 40, 42; Herts. RO, Moulton Coll. 46325; my *Ag. Rev.*, 184sqq.; 'Reconsideration', 31.

31 Ibid. 32; *Fmrs*, pl. 12; Blith, op. cit., 103; BL, Harl. MS 3749, fos 13, 14.

32 Bateson et al., op. cit., xii, 239; Vivian, op. cit., 1, 27, 200–1, 203–4; Straton, op. cit., 113; my 'Reconsideration', 32; *Ag. Rev.*, 184sqq.; W. J. Slack, *Ldp Oswestry*, Shrewsbury, 1951, pp. 56, 77; E. Straker, *Wealden Iron*,

1931, p. 299; Leconfield (Ld), *Petworth Man. in 17th cent.* 1954, pp. 64–5; [Hervey, Ld. Francis], *Ickworth Surv. Booke* s.l. n.d. [1893], pp. 8, 10, 11; H. Fishwick, *Surv. Man. Rochdale 1626*, Chetham Soc., n.s., 1913, lxxi, 28, 37–40, 42, 74, 179, 183, 197–9, 220, 222, 241; W. P. W. Phillimore and G. S. Fry, *Abs. Glos. Inqs P. M. in reign K. Chas I*, Ind. Lib., Brit. Rec. Soc. 3 pts (vols) 1893, i, 130; R. Stewart-Brown, *Ches. Inqs P.M., Stuart Per. 1603–60*, 3 vols, Rec. Soc. f. Lancs. and Ches. 1934–8, lxxxiv, lxxxvi, xci, vol. ii, 2, 3; iii, 85; K. M. Dodd, *Fld Bk Walsham-le-Willows 1577*, Sfk Recs Soc. 1974 xvii, 56, 65, 81, 102–3, 105, 107–8, 110, 112, 118, 124, 134, 143–4, 152; F. N. Fisher, 'N. on Man.Hist. Horsley', *Derbys. Archaeol. Jnl* 1952 xxv, 49; A. R. H. Baker, 'Fld Systs in V. Holmesdale', *Ag. Hist. Rev.* 1966 xiv, 10; PRO, Ct Wards and Liveries, Feodaries' Survs, bdl. 30 pt ii no. 702; St. Ch., Jas 16/4 m. 3; SPD Jas 8/76, fo. 5; 195/18 fo. 113; D.L., M.B. 113 fo. 3v; R.&S.bdl. 2 no. 2; Sp. Colls, R.&S., G.S. roll 986; portf. 1 no. 28; Exch., A.O., Parl. Survs: Corn. 40 fo. 3; Ess. 12 fo. 2; Hants. 14 fos 1–3; Herefs. 19 fos 4, 11; Herts. 7 fos 7, 8; Kent 26 fos 3, 4; Lancs. 18 fos 25sqq.; Lincs. 10A fo. 9: 16 fos 14, 15, 17; 20 fo. 66; Northants. 15 fo. 3; 32 fo. 57(1); 33 fo. 3; Oxon. 10 fo. 2; Surrey 32 fo. 4; 60 fo. 4; 67 fos 3, 4; Wmld 1 fo. 4; Worcs. 2 fos 2, 3; 6 fo. 4; Yks 27 fo. 5; 35 fos 9, 11; 39 fo. 10; K.R., M.B. 40 fos 24sqq.; Deps by Commn 32 Eliz. East. 10 ex parte quer.; L.R., M.B. 185 fos 2 (1), 15 (14), 88v, 100v, 102, 134(35)v, 156(57); 186 fos 92v, 94; 210 fo. 40; 218 fo. 81; 219 fo. 322; 228 fo. 165(17); 230 fo. 72; 258 fo. 125; BL, Add. MSS 14850 fo. 153v; 27605 fos 7, 106v; Stowe MS 858 fo. 62; Harl. MSS 6697 fos 2–4; 7369 fo. 19; Ess. RO, D/DP E26 fo. 84; D/DP M186 pp. 25, 28–9, 33, 51–2; Worcs. RO, Acc. 494 part. Morton Underhill 1648; Shrops. RO, MS 167/43; Birm. Ref. Lib. MS 382959 fo. 9v; 478550; Soms. RO, DD/CN Acc. C. 168 box 2 no. 26 p. 95; DD/X/GB Acc. W9 Combe Surv. Bk pp. 127, 175, 228; Dors. RO, D.54 Terrier w. rental Ryme Intrinsecus 1563; Sfk RO (Ips.) 51.10.17.3 fo. 39; Lincs. RO, Heneage of Hainton Coll. 3/3; Ches. RO, Nedeham and Kilmorey Coll. surv. n.d. The Hack and Bromall; Deene Ho. Brudenell Coll. O.x.14; Herts. RO, Ashridge Coll. 232; Gor'bury Coll. XI.11 fo. 60; Glos. RO, D.184 M24 surv. bk p. 203; Lancs. RO, DD.K 1451/1 mm. 8, 26; 1452/1 mm. (1st occ.) 26, 28, (2nd occ.) 27, 32, 34, 37; 1456/2 m. 4 (1st occ.); 1465/1 m. 29; 1470/1 m. 20.

33 Mead, art. cit., bet. pp. 40–1.

34 Boughton Ho. Compota Omnium et Singularum Officiorum et Ministorum f. 2 y.e. 38 Hy 8, fo. 2v; Deene Ho., Brudenell MSS A.iv.1; K. M. Dodd, op. cit., 55, 63, 69, 73–4, 76, 79, 80, 83, 92–3, 109, 111–12, 114–17, 120, 122–3, 135–8, 140–1, 147, 152; Baker, 'Fld Systs Holmesdale', 12; Marshall, *Nfk*, i, 130; [Hervey], *Ickworth*, 15, 16, 19, 56; Brist. RO, MS 4490 survs 1542–7 fo. 13v; Dors. RO, D.54 Terrier w. rental Melbury Osmond 1563; Worcs. RO, B.A. 68 cov. memo. Woodmanton le. 3 Apr. 1612; B.A. 494 part. Morton Underhill 1648; PRO, Exch. L.R., M.B. 228 fo. 151(3); A.O., Parl. Survs: Worcs. 6 fo. 3; Yks 34 fo. 3; Ess. RO, D/DP M186 p. 54; M890 p. 13; Sfk RO (Ips.) 51.2.12 fo. 25v; V5.23.2.1 fo. 17; BL, Add. MS 34162 fo. 40; Northants. RO, Mont. Coll. box 1322 part. Man. Beaulieu 1832, p. 29.

35 Ibid., Misc. Led. 45 pp. 184, 451; BL, Harl. MS 5827 fo. 7(5)v; Norden, *Surveiors Dialogue*, 229; A. W. Boyd, *Country Par.*, 1951, p. 7.

36 E.g. my *Ag. Rev.*, 99, 100; *Fmrs*, pl. 11; 'Ridge and Furrow', 19, 24; 'Reconsideration', 32sqq.; W.J. Corbett, 'Eliz. Vil. Survs', *Trans. R. Hist. S.*, n.s. 1897 xi, 86–7; A.C. Chibnall, *Sherington: fiefs and flds Bucks. Vil.*, Cambridge, 1965, pp. 235, 287; D.M. Barratt, *Eccl. Terriers War. Pars*, 2 vols, Dugdale Soc. xxii, xxvii, 1955–71, ii, 118; D.C. Douglas, *Soc. Str. Med. E. A.*, Oxf. Studs in Soc. and Leg. Hist. ix, Oxford, 1927, p. 18; Blith, op. cit., 80, 100; H.S. Bennett, *Life on Engl. Man.*, Cambridge, 1938, p. 45; R. Douch, 'Customs and Trads I. Portland', *Antiquity* 1949 xxiii, 152; A.J. Davison, 'Some Aspects Agrarian Hist. Hougham and Snetterton as revealed in Buxton MSS', *Nfk Archaeol.* 1973 xxxv, 349; R.C. Hoare, *Hist. Mod. Wilts.* Hund. Warminster, 81–2; Marshall, *Midl. Cos*, ii, 225; *Midl. Dept*, 603; *E. Dept*, 135–6; *Desc. Cat. Man. R. bel. to Sir H.F. Burke*, Man. Soc. Pubs, xii, 1923, pt ii, p. 22; Kalm, op. cit., 215; Morton, op. cit., 14; H.L. Gray, *Engl. Fld Systs*, Cambridge, Mass., 1915, pp. 174, 227–9; W. Cooper, *Wootton Wawen: its hist. and recs*, Leeds, 1936, app. p. 12; F.S. Colman, *Hist. Barwick-in-Elmet* Thoresby Soc. 1908 xvii, 127; Bateson et al., op. cit., ix, 119, 194; x, 270; *V. C. H. Glos.*, ii, 164; C.M. Hoare, *Hist. an E.A. Soke*, Bedford, 1918, p. 249; E.C.K. Gonner, *Com. Ld and Inclo.*, 1912, p. 27; H.T. Crofton, 'Rels Com. Fld Syst. in and nr Man.', *Man. Qtly* Jan. 1887, pp. 9, 11; T. Lawson-Tancred, '3 17th cent. Ct R. Man. Aldborough', *Y. A. J.* 1943 xxxv, 202; M. Davies, 'Fld Patts in V. Glam.', *Cardiff Naturalists' Soc. Reps and Trans.* 1954–5 lxxxiv, 8, 11; 'Open Flds Laugharne', *Geog.* 1955, p. 176; 'Rhosili Open Fld and rel. S.W. Fld Patts', *Ag. Hist. Rev.* 1956 iv, 94–6; M.R. Postgate, 'Fld Systs Breckland', ibid. 1962 x, 83; Worlidge, op. cit., 273; Ault, *Husb.* 36, 87; A.G. Ruston and D. Witney, *Hooton Pagnell*, 1934, p. 61; J. Waylen, *Hist. Mil. and Mun. Tn Marlborough* 1854, p. 120.

37 My 'Reconsideration', 38–9; Gonner, op. cit., 27; N.J. Hone, *Man. and Man. Recs*, 1912, p. 191; F.R. Goodman, *Rev. Landlds and their Tens*, Winch. 1930, p. 50; W.M. Palmer, *Hist. Par. Borough Green, Cams.* C. A. S. Oct. Pubs 1939 liv, 154; G. Slater, *Engl. Peasantry and Enclo. Com. Flds*, 1907, pp. 21, 43; Shak. Bpl. Trevelyan Coll. Ct R. Snitterfield and Bearley 13 Oct. 1703, 27 Oct. 1713 (1715 erron.), 26 Oct. 1715; Oxon. RO, Dashwood Coll. VIII/xxxiii Ct R. Duns Tew 30 Sept. 13 Chas 2; Ess. RO, D/DVm 20 fo. 33; Wilton Ho. Ct R. 1689–1723 box ii, vol. ii, p. 18; PRO, Ct R., G.S. bdl. 195 file 78 m. [3d.]; Lancs. RO, DDF. 168 Ct R. Leyland 1 May 1691; Beds. RO, BS. 1276 Ct R. Tempsford 30 Apr. 1617; L. 26/563 Ct R. Harrold 2 May 9 Chas; TW. 10/2/9 ct files Clapham: by-laws 14 Apr. 8 Chas; Bucks. RO, IM 6/1 Ct R. Ivinghoe 30 Sept. 9 Jas; BL, Add. MS 27977 fo. 200v; Add. R. 27171; Northants. RO, Ecton Coll. 1183, 1185, 1189, 1191; Daventry Coll. 585 Barby cum Onely ords 23 Oct. 1730; Mont. Coll. Northants. old box 18 no. 160 Ct Bk Barnwell ords 9 Dec. 5 Chas; Misc. Led. 145 p. 510.

38 Ibid. Aynho Ct. R. 9 Apr. 1680; PRO, Exch. KR Deps by Commn 26 Eliz. Trin. 8 dep. Wm Harrowden; H. Stocks, *Recs Boro. Leicester 1603–88*, Cambridge, 1923, pp. 201–2.

39 F. Seebohm, *Engl. Vil. Commun.*, 1883, pp. 4, 19, 20; Marshall, *Midl. Dept*, 466; my 'Reconsideration', 39, 40; Postgate, 'Fld Systs', 99; Crofton, art. cit., 8, 9, 11; Corbett, art. cit., 86; cf. M. Spufford, 'Rural Cambs. 1520–1680', ts. thes. M.A. Leicester Univ. 1962, pp. 95sqq.

40 Corbett, art. cit., 72; Gonner, op. cit., 27; E. D. R. Burrell, 'Hist. Geog. Sandlings Sfk, 1600–1850', ts. thes. M.Sc. London Univ. 1960, p. 46; my 'Reconsideration', 39, 40.

41 Wilts. RO, Enclo. Acts and Awards, Foffont etc. 1785, 1792, Award, p. 49.

42 My *Farmers*, pl. 11; Hallam, *Set. and Soc.*, 140, 142, 150–3, 171; Kalm, op. cit., 261–2; N. Neilson, *Terrier Fleet* in Brit. Ac. Recs Soc. and Econ. Hist. E. W. vol. iv, 1920, p. lxv; Marshall, *Midl. Dept* 466; Ault, *Husb.*, 93; H. S. Bennett, op. cit., 48; *V. C. H. Glos.*, ii, 164; Leics. RO, Mus. MS 10D.41 Ct Bk Ullesthorpe 1 Oct. 1690; Northants. RO, F-H Coll. 272 fo. 83f; 541 m. 2; 834 Gretton by-laws 24 Oct. 1695; I(L) Coll. 128 Ct. R. Lamport Apr. 21 Eliz., 1 Oct. 2 Jas; Mont. Coll. Northants. old box 17 no. 160 Ct Bk Broughton 3 Mar. 15 Eliz.; old box 7 no. 66/10 Ct R. Weekley 12 Oct. 1721; Misc. Led. 129 Ct Bk 1721–8, Weekley 12 Oct. 1721; Shak. Bpl. Leigh Coll. Ct R. Ratley 22 Sept. 1574; BL, Add. R. 47373; Add. MSS 23150 fo. 23; 36875 fo. 31; 36903 fo. 209; PRO, DL, Ct R. 81/1101 Cas. Donington 7 Oct. 6 Eliz.; 1133 m. 28.

43 Wilts. RO, Enclo. Acts and Awards, Foffont etc. 1785, 1792, Award, pp. 141–2.

44 Chetham Lib. Mun. Rm, Adlington MSS fo. 169; Sfk RO (Ips.) V.5.23.2.1 fo. 17; Ess. RO, D/DVm 20 fo. 36v; K. M. Dodd, op. cit., 157; Corbett, art. cit., 86–7; G. A. Thornton, *Hist. Clare*, Cambridge, 1930, p. 112; J. A. Venn, *Founds Ag. Econ.*, Cambridge, 1933, p. 44, f.p. 64.

45 Ibid. 32; Slater, *Engl. Peasantry*, 21; Seebohm, op. cit., 5, 6; P. Vinogradoff, *Engl. Soc. in 11th cent.*, Oxford, 1908, p. 279; Douch, art. cit., 152; my 'Reconsideration', 34–6; J. Aubrey, *Nat. Hist. Wilts.* ed. J. Britton, 1847, p. 32; A. G. Street, *Ditchampton Fm*, 1946, p. 14; G. Atwell, *Faithfull Surveyour*, Cambridge, 1662, p. 90. Cf. J. W. Macnab, 'Brit. Strip Lynchets', *Antiquity* 1965 xxxix, 279sqq.; H. C. Bowen, *Anc. Flds: tentative anal. vanishing earthworks and landscapes* B.A.A.S. 1961, pp. 3, 16, 17, 25, 34, pls iv, v.

46 Cobbett, op. cit., ii, 81; Marshall, *S. Cos*, ii, 301–2.

47 BL, Add. MS 15350 fo. 62 (63)v; Wilts. RO, Acc. 283 Liber Supervisus maneriorum de Amisburie Erles et Amesbury Pryorye; Amesbury Surv. Bk 1574; Surv. Amesbury 1635 fos 14, 47–8.

48 *Cal. Proc. Chanc. in reign Q. Eliz.*, 3 vols, Rec. Commn 1827–32, i, 60; A. H. Shorter, 'Anc. Flds in Manaton Par. Dartmoor', *Antiquity* 1938 xii, 183sqq.; M. Hartley and J. Ingilby, *Yks. Vil.*, 1953, p. 238; A. Raistrick, *Malham and Malham Moor*, Clapham via Lanc. 1947, pp. 54–5, 68; *V. C. H. Glos.*, ii, 164–5; Harvey and St Leger-Gordon, op. cit., f.p. 63, p. 150; Seebohm, op. cit., 6; D. J. Robinson, J. Salt and A. D. M. Phillips, 'Strip Lynchets in Peak Dist.', *N. Staffs. Jnl Fld Studs*, 1969 ix, 92sqq.; H. L. Gray, op. cit., 377; Hennell, op. cit., 63; Rider-Haggard, op. cit., ii, 98, 158; Douglas, op. cit., 18; Venn, op. cit., 32.

49 Ibid.; Douch, art. cit., 152; my *Fmrs*, pls 2, 17.

50 PRO, P.C.R. 1631–2 fos 253v–4 (pp. 506–7); Leic. Tn Hall MS BR. II.8.75; BL, Egerton MS 3622 fos 201v–3; Stark, op. cit., 93; Eland, *Purefoy Ltrs* ii, 438; A. C. Chibnall, op. cit., 235, 287; my *Ag. Rev.*, 100sqq., 107–8.

51 BL, Egerton MS 3134 fos. 146sqq.

52 Glos, RO, D.184 M24 pp. 135sqq.; Shak. Bpl. Man. Docs, Ct R. Alveston and Tiddington 10 Oct. 1740; W. de B. Coll. 1253; Barratt, op. cit., i, 48; Phillimore and Fry, op. cit., i, 44–5, 105–7.
53 Crofton, 'Relics', 8, 9, 11; BL, Add. MS 36920 fo. 37; Staffs. RO, D. 1750 Ct R. Madeley Holme 11 Oct. 1655.
54 My *Ag. Rev.*, 117, 123, 142–5, 159, 174.
55 Shak. Bpl. Leigh Coll. Ct R. Ratley 20 Oct. 1602; Ct Bk Adlestrop fo. 10; PRO, D.L., M.B. 117 fos 75–6; BL, Add. MSS 23151 fo. 55; 36585 fos 163v, 167; Birm. Ref. Lib. 168162, 7 Oct. 1700; 168163, 12 Oct. 1702; Glos. RO, D184 M7 nos 1, 2; Oxon. RO, Dashwood Coll. VIII/xxxiii Ct R. Duns Tew 5 Apr. 15 Jas, 30 Sept. 13 Chas 2.
56 G. S. and E. A. Fry, *Abs. Wilts. Inqs P. M. temp. Car. I* Ind. Lib., B.R.S. 1901, pp. 92–3; PRO, Wds, Feodaries' Survs, Wilts. bdl. 46, unstd survs: Jn Bryan(t) 1620 S. Marston; Exch. L.R., M.B. 191 fo. 156; BL, Add. MS 37270 fo. 121.
57 Ibid. fo. 114; cf. Bowood Ho., Surv. Bremhill 1629.
58 Straton, op. cit., 156, 285–6, 291–4; Hoare, op. cit., Hund. S. Domerham pp. 47, 52–3; G. E. Fussell, *Robt Loder's Fm Accts 1610–1620* R. Hist. S., Camd. 3rd ser. 1936 liii, 11, 12; F. J. Baigent and J. E. Millard, *Hist. And. Tn and Man. Basingstoke* Basingstoke, 1889, p. 117; E. J. Bodington, 'Ch. Survs in Wilts. 1649–50', *W. A. M.* xl, xli, 1917–22, xli, 30; T. S. Maskelyne and F. H. Manley, 'N. on Eccl. Hist. Wroughton', ibid., 476; PRO, Exch. K.R., Sp. Commn 2422; LR, MB 191 fos 152–3; 197 fo. 107; Chanc. Proc. ser. i Jas M.13/55; Wilton Ho., Survs Manors 1631 i, Fugglestone fo. 8 (9); Surv. Flambston 1631 fos 2, 4, 6.
59 PRO, Exch. L.R., M.B. 197 fo. 107; Berks. RO, D/EC M26; N. S. B. and E. C. Gras, *Econ. and Soc. Hist. Engl. Vil.*, Cambridge, Mass. 1930, pp. 592–3.
60 Wilts. RO, Savernake Coll. surv. Burbages 1574 fo. 26.
61 Wilts. RO, Acc. 212B, BH. 8, Broad Hinton rental 1636.
62 Straton, op. cit., 311–12.
63 PRO, Wds, Feodaries' Survs, Wilts. bdl. 46, unstd survs: Baskervile conf. 19 Jas; St. Ch., Eliz. A.8/37; A.11/8, dep. Jn Mattocke; W. Money, *Purveyance R. Hsehld in Eliz. Age, 1575*, Newbury, 1901, pp. 111–12; Maskelyne and Manley, art. cit., 476.
64 Wilts. RO, Savernake Coll. Surv. Collingbourne Kingston 1595 fo. 54.
65 G. W. Kitchin, *Man. Manydown* Hants. Rec. Soc. 1895 x, 184–5.
66 My *Ag. Rev.*, 60.
67 M. Spufford, *Cams. Commun.*, Leicester, 1965, p. 43; BL, Add. MS 14850 fo. 145; PRO, Exch. L.R., M.B. 201 fos 56v, 58; Marshall, *Nfk*, ii, 384.
68 C.C.C.C. MS 173 fo. 50r repro. in R. Flower and H. Smith, *Parker Chron. and Laws* E.E.T.S. 1941; cf. Pub. Rec. Commnrs, *Anc. Laws and Insts Engld* 1840, p. 55–6; pace W. Kirbis, *Siedlungs- und Flurformen german. Länder, bes. Grossbritanniens, im Lichte der dtsch. Siedlungsforschung* Göttinger Geogr. Abh. Heft 10, 1952, p. 45; H. P. R. Finberg in Finberg and J. Thirsk, *Agrarian Hist. E. W.*, sev. vols, Cambridge, 1967– in prog., i, pt ii, pp. 416–17; H. S. A. Fox, 'Approaches to Adoption Midl. Syst.' in T. Rowley, *Origs Open-Fld Ag.*, 1981, p. 87. For fences, see e.g. G. C. Homans, *Engl. Villagers 13th cent.*,

Cambridge, Mass. 1941, pp. 65–6; P. Vinogradoff, *Growth Manor*, 1904, p. 261; Ault, *Husb.*, 30, 60–1; *Open-Fld Fmg in Med. Engld: study in vil. by-laws*, 1972 pp. 52, 91, 93; Wilton Ho. Ct R. Mans 1689–1723, box iii, vol. iv, p. 46; Ches. RO, Vernon (Warren) Coll. Ct R. nos 2, 14; Herts. RO, Moulton Coll. 44467 Ct R. Pirton 1696; Oxon RO, DIL.II/w/75, 13 Apr. 16 Chas; PRO, D.L., Ct R. 106/1535 mm. 5, 6d., 13; Sp. Colls, Ct R., G.S. 195/36 mm. 1d.–3; 195/78 m. [1]; 79 m. 6d.; Staffs. RO, Hand Morgan Coll. Chetwynd MSS, file N, Ct R. Churcheaton 23 Oct. 41 Eliz.; file O, Ct R. Ingestre and Salt 31 Mar. 23 Eliz.

69 H. L. Gray, op. cit., 61–2; Straton, op. cit., p. lxv; Kerridge, *Survs*, 35, 37, 39, 104, 131, 136–7; W. Brown, *Yks. Deeds*, 3 vols, Yks. Archaeol. Soc. Rec. Ser. xxxix, l, lxiii 1909–22, i, 165–6; P. Millican, *Hist. Horstead and Stanninghall*, Norwich, 1937, pp. 34, 84.

70 BL, Harl. MS 4660 fo. 8 (11).

71 Soc. Antiquaries Lond. MS 60. fo. 47 in A. J. Robertson, *Anglo-Sax. Chs*, Cambridge, 1956, pp. 78, 80; cf. pp. 81, 331.

72 BL, Stowe Ch. 36.

73 BL, Cott. Tib. A.xiii fo. 66; also Vesp. A.v fo. 188v.

74 BL, Cott. Tib. A.xiii fo. 74.

75 Ibid. fo. 94v.; and see Harl. MS 4660 fo. 8 (11).

76 As n. 71.

77 BL, Cott. Tib. A.xiii fo. 84v; and see Harl. MS 4660 fo. 9 (13) (Moreton in Bredon).

78 BL, Cott. Tib. A.xiii fo. 96v.

79 Bede, *Historiae Ecclesiasticae Gentis Anglorum* ed. J. Smith, Cam. 1722, app. xxi, p. 779.

80 BL, Cott. Claud.B.vi fos 93v–4.

81 BL, Add. MS 15350 fo. 47 (49).

82 BL, Cott. Claud.B.vi fo. 86 (r et v); and see fo. 85v.

83 Ibid. fo. 80v.

84 Ibid. fos 61v., 94 (r et v).

85 As n. 90.

86 Cf. D. Hooke, 'Open-Fld Ag.: evid. fr. pre-conquest chs W. Midlds' in Rowley, op. cit., p. 58.

87 BL, Cott. Claud.B.vi fo. 40v.

88 Ibid. fo. 101 (102); C.ix fo. 127 (125) (23)v; Aug. ii fo. 48.

89 BL, Add. MS 15350 fo. 105 (106).

90 BL, Harl. MS 596 fo. 17; and see Straton, op. cit., 265–8; my *Ag. Rev.*, 16–19; C. R. Hart, *Early Chs E. Engld*, Leicester, 1966, pp. 253–4; as n. 79.

91 BL, Cott. Claud.B.vi fo. 75.

92 BL, Harl. MS 436 fo. 65; Cott. Aug.ii, fo. 76.

93 Pub. Rec. Cmmnrs, *Anc. Laws and Insts*, 185–6, 188, and esp. 189. Cf. P. Vinogradoff, *Villainage in Engld: essays in Engl. med. hist.*, Oxford, 1892, p. 356; Homans, op. cit., 293sqq.; H. S. Bennett, op. cit., 179–80.

94 E. O. Blake, *Liber Eliensis* R. Hist. S., Camd. 3rd ser. 1962 xcii, 111; W. de G. Birch, *Cartularium Saxonicum*, 3 vols, 1885–93, pp. 227sqq., 347–8; E. King, 'Este Recs Hotot Fam.' in his *Northants. Miscellany* Northants. Rec. Soc. 1983 (1982) xxxii, 31; W. O. Hassall, *Wheatley Recs. 956–1956* Oxon.

Rec. Soc. 1956, pp. 27–8; F. M. Stenton, *Docs. Illus. Soc. and Econ. Hist. Danelaw*, 1920, pp. 20–1, 140–1, 170–2, 385–6; E. Mason, *Beauchamp Cartulary: Chs 1100–1268* Pipe R. Soc. n.s. 1980 (1971–3) xliii, 14, 16, 17, 80–1, 87–8, 123, 125, 129–30, 135–6, 157–9; R. R. Darlington, *Cartulary Worc. Cath. Priory (Reg. I)* Pipe R. Soc. 1968 (1962–3) lxxvi, 13; *Cartulary Derby Abb.*, 2 vols, p.p. 1945, pp. 102–3, 246–7, 255–8, 313–14, 529–32; J. E. Jackson, *Liber Henrici de Soliaco Abbatis Glaston. et vocatur A: Inq. Mans Glaston. Abb. in yr MCLXXXIX* Rox. Club 1882, pp. 12, 29, 39, 42, 105, 114–15, 137, 140, 155 et pass.; as nn. 70, 75, 79; D. Hooke, 'Early Forms Open-Fld Ag. in Engld', *Geografiska Annaler* 1988, lxxB, pp. 124–6, and 'Early Med. Estes and Set. Patts: doc.evid.' in M. Aston, D. Austin and C. Dyer, *Rural Sets Med. Engld: studs ded. to M. Beresford and J. Hurst*, Oxford, 1989, pp. 21–2, has recently mapped some early flds. Cf. A. Campbell, *Chs Roffen* 1973, pp. 45, 47.

95 J. M. Kemble, *Codex Diplomaticus Aevi Saxonici*, 6 vols, 1839–48, i, 38sqq., 216–17, 232–3; ii, 1, 2, 57–8, 100–1, 243–4; iv, 201–3; v, 276–7, 318–20; vi, 115–16, 214; T. Madox, *Formulare Anglicanum, or Coll. Anc. Chs and Instrs*, 1702, pp. 134–5; Birch, op. cit., 217, 227–31; Cf. Finberg and Thirsk, op. cit., i, pt ii, 493. For hitching, M. Chibnall, *Sel. Docs Engl. Lds Abb. Bec* R. Hist. S. Camd. 3rd ser. 1951 lxxiii, 90, 135; Ault, *Fmg*, 82–3; *Husb.* 56; J. E. Jackson, op. cit., 12, 155.

96 D. Whitelock, *A.-S. Wills*, Cambridge, 1930, pp. 72, 74, 76, 84, 86, 88, 94. For Brandon, Blake, op. cit., 111. For acres, e.g., D. Sylvester, *Rural Ldscape W. Borderlds: stud. in hist. geog.*, 1969, p. 307.

97 Douglas, op. cit., 18–21, 23, 28, 229; cf. F. M. Stenton, *A.-S. Engld*, Oxford, 1971, p. 313.

98 Bateson et al., op. cit., vii, 316; x, 133, 270, 273; my *Ag. Rev.*, 166; Ault, *Husb.*, 32; J. Thirsk, 'Com. Flds' in R. H. Hilton, *Peasants, Knts and Heretics: studs in med. Engl. soc. hist.*, Cambridge, 1981, p. 24.

99 Ibid., 23–4; A. R. H. Baker, 'Some Flds and Fms in Med. Kent', *Arch. Cant.* 1965 lxxx, 168–9; W. Austin, *Hist. Luton*, 2 vols, Newport, I.W. 1928, i, 259–60; ii, 271–3; R. J. Whiteman, *Hexton: par. surv.* p.p. 1936, pp. 99, 101, 103; BL, Add. R. 27169, 27173; Add. MS 27977, fos 3, 358; Bucks. RO, Ct R. Pitstone P. 24/1, 9 Apr. 9 Jas, 5 May 15 Jas; P. 24/2, 7 Oct. 7 Chas; P. 24/3, 14 Apr. 1651, 12 Oct. 1677; IM.6/12 Ct Bk Ivinghoe, p. 2; Herts. RO, Ashridge Coll. 656, 677, 683, 723, 841–4; 2667, 25 Sept. 22 Jas, 19 Jan. 8 Chas; Pym Coll. 19271, Ct R. Norton 28 Apr. 1606, 11 Apr. 1608; Moulton Coll. 44467, Ct R. 1696; Cashiobury and Gape Coll. 9442; 9464, 29 Mar. 17 Jas; Misc. MSS 7052, 1 Dec. 5 Chas; 41674, fo. 13; 49151; 65810, pp. 143sqq.

100 Marshall, *S. Dept*, 8; D. Walker, *Gen. View Ag. Co. Hertford*, 1795, pp. 49, 50; Aspley Hth Sch. Hist. Soc., *Hist. our Dist.*, Aspley Guise, 1931, pp. 33–5; J. C. Wilkerson, *Jn Norden's Surv. Barley, Herts. 1593–1603* Cam. Ant. Recs Soc. 1973 ii, 32sqq.

101 My *Ag. Rev.*, 129, 145.

102 H. S. Bennett, op. cit., 49; as n. 422; Ault, *Husb.*, 30.

103 Ibid., 12, 48–9, 54; S. C. Powell, *Puritan Vil.*, Middletown, Conn., 1963, pp. 16, 17.

104 As nn. 429–30.

105 *Desc. Cat.* pt i, p. 19; pt ii, p. 13; Man. Lib., Misc. Dd M.7; Mx RO, Acc. 249/875, c. customs Ruislip 1640; Birm. Lib. MS 437912; Staffs. RO, Hatherton MSS box 16 bdl. (b) Surv. Penkridge 1660 fo. 23; PRO, Exch. A.O., Parl. Surv. Sfk 19 fo. 10; Lancs. RO, DD/K 1505/11.

106 As nn. 421–30.

107 Ault, *Husb.* 26; Vinogradoff, *Engl. Soc.*, 389; my *Ag. Rev.*, 43–6, 52, 58sqq., 71sqq., 96–7, 100–1, 108, 113–14, 117, 119, 121, 143, 149, 174, 311–13; T. Tusser, *500 Pts Gd Husb.* ed. G. Grigson, Oxford, 1984, pp. 99, 277; cf. R. Lennard, *Rural Engld 1086–1135*, Oxford 1959, pp. 263–4.

108 Seebohm, op. cit., 120sqq.; G. G. Coulton, *Med. Vil.*, Cambridge, 1925, pp. 40–2, 90; T. Lewis, 'Seebohm's Tribal Syst. of W.', *Econ. Hist. Rev.* 2nd ser. 1956 ix, 31–3; Baker, 'Flds and Fms', 167–8.

109 Wilts. RO, Ex Sarum Dioc. Reg., Misc. Wills, Admons and Invs, Wr Kember 1585 Marten in Gt Bedwyn; Dorothy Goodyer 1633 Mere; Cons. Ct Sarum inv. Jn Earle 1615 Patney; Leics. RO, Arch. invs 1636/132; 1693/110; Lichfield Jt RO, Pecs, invs, Robt Linnes 1634 Blithbury; Th. Cowper 1637 Pipe Ridware; M. A. Havinden, *Hsehld and Fm Invs in Oxon. 1550–90* H. M. C. Jt Pub 10 (Oxon. Rec. Soc. xliv) 1965, pp. 284–5. Pace W. G. Hoskins, *Essays in Leics. Hist.*, Liverpool, 1950, p. 149; A. R. H. Baker, 'Open Flds and Partible Inheritance on Kent Man.', *Econ. Hist. Rev.* 2nd ser. 1964 xvii, 17, 22–3.

110 Marshall, *Glos.*, i, 76–7; Leconfield, Ld, *Sutton and Duncton Mans*, 1956, pp. 61–2; Vinogradoff, *Villainage*, 253; Homans, op. cit., 78–9, 81; H. S. Bennett, op. cit., 45–6; Ault, *Husb.*, 31–2; Pub. Rec. Cmmnrs, *Anc. Laws and Insts*, 61; BL, Cott. MS Nero, A.i, fo. 70 (71); Lewis, 'Seebohm's Tribal Syst.', 31–2.

111 Ibid.; E. Lloyd, *Hist. Wales*, 2 vols, 1939, i, 296, 318–19; D. McCourt, 'Infld and Outfld in Ire.', *Econ. Hist. Rev.* 2nd ser. 1955 vii, 375–6.

112 Blith, op. cit., 28; Atwell, op. cit., 88–9; Mortimer, op. cit., 41; S. and B. Webb, *Engl. Loc. Govt: Man. and Boro.* pt i, 1908, p. 79; W. G. Fletcher, *Hist. Loughborough*, Loughborough, 1887, p. 46; PRO, D.L., Ct R. 106/1535 m. 5d.; Cams. RO, L.1/112, Oct. 1654; Beds. RO, HA.5/1 fo. 22v; Northants. RO, Mont. Coll. Northants. old box 18 no. 160 Ct Bk Barnwell 1 Dec. 1614, 1624, 9 Dec. 5 Chas.; sep. sht, 17th cent. ords; Hunts. old box 26 no. 28 Ct R. Caldecot Dec. 5 Chas.; Wmld Coll. 5.v.1 Ct R. Stanground 3 Sept. 7 Jas, 16 Sept. 1624; Farcet 21 Oct. 1660; Woodston 16 Sept. 12 Jas, 12 Sept. 1623, 16 Sept. 1624; Old Par. Recs, Fld offrs' accts 1738; Misc. Led. 129 Caldecot Oct. 1721. Pace Hoskins, *Essays*, 149; *Midl. Peasant*, 1957, pp. 159, 241, 245.

113 Mortimer, op. cit., 41; Cams. RO, L.88/4 Ct R. Caxton, presentments Whitsun 1659, 4 Oct. 1666.

114 J. Mastin, *Hist. and Ants Naseby*, Cambridge, 1792, p. 18.

115 C. S. and C. S. Orwin, *Open Flds*, Oxford, 1938, pp. 5, 6, 8, 9, 40–1.

116 My *Ag. Rev.*, 150–1, 157, 159, 161–2, 167, 172; Bateson et al., op. cit., ii, 478; v, 424, 488; ix, 324; xiv, 212; J. Raine, *Priory Hexham*, 2 vols, Surtees Soc. xliv, xlvi, 1864–5, ii, 50; Finberg in Finberg and Thirsk, op. cit., i, pt ii, 495; as n. 88.

117 G. M. Cooper, 'Berwick Par. Recs', *Suss. A. C.* 1853 vi, 239; Ault, *Fmg*, 27; R. A. Dodgshon, *Origs Brit. Fld Systs: interpretation*, London and New

York, 1980, pp. 34–5; Bracton, *On Laws and Customs of Engld: De Legibus et Consuetudinibus Angliae* ed. G. E. Woodbine and S. E. Thorne, 4 vols, Cambridge, Mass. 1968–77, ii, 220; G. Elliott, 'Enclo. Aspatria', *Trans. Cumb. and Wmld Ant. and Archaeol. Soc.* n.s. 1960 lx, 100; PRO, Exch. L.R., M.B. 222 fo. 293; H. S. Bennett, op. cit., 55.

118 J. D. Chambers and G. E. Mingay, *Ag. Rev. 1750–1880*, 1966, p. 51.

119 Wilton Ho. Survs Mans 1631 vol. i, fos 11 (12), [12] (13), marginalia.

120 M. C. Naish, 'Ag. Landscape Hants. Chalklands, 1700–1840', ts. thes. M.A., London Univ. 1960, p. 142.

121 Hoare, op. cit., Hund. Chalk, pp. 153–4.

122 Urchfont Manor Ho., copies enclo. act and award 1789, 1793; Wilts. RO, Enclo. Acts and Awards, Foffont etc. 1785, 1792; Wilton Ho. Survs Mans 1631 vol. i, Broad Chalke; vol. ii, Stoke Farthing (Verdon); Ct Bk id. beg. 1743, entries 1794, 1799; Surv. Man. Alvediston 1706 and 1758, pp. 1, 4–6; T. Davis, *Gen. View Ag. Co. Wilts.*, 1813, p. 43.

123 Ibid., 1794 edn, p. 15; Wilts. RO and Urchfont Man. Ho.: as prev. n.; Marshall, *Yks.*, i, 51.

124 Fishwick, op. cit., 106, 119, 126; Crofton, 'Relics', 9; A. N. Palmer and E. Owen, *Hist. Anc. Tenures Ld in N. W. and Marches*, p.p. 1910, p. 3; my *Ag. Rev.*, 166; Soms. RO, DD/CN Acc. C.168 box 2 no. 26 p. 95.

125 T. A. M. Bishop, 'Assarting and Growth Open Flds' in E. M. Carus-Wilson, *Essays in Econ. Hist.*, vol. [i] 1954, pp. 26sqq.; B. M. S. Campbell, 'Pop. Chg. and Gen. Com. Flds in E. A.', *Econ. Hist. Rev.* 2nd ser. 1980 xxxiii, 181; cf. *V. C. H. Leics.*, ii, 158.

126 D. Willis, *Este Bk Hy de Bray of Harlestone, Co. Northants. (c. 1289–1340)* R. Hist. S., Camd. 3rd ser. 1916 xxvii, 86–8.

127 Stenton, *A.-S. Engld*, 313; Ault, *Fmg*, 129; Kerridge, *Survs*, 14, 18, 88, 91–4, 102, 111; Bateson et al., op. cit., x, 270; B. R. Kemp, *Reading Abb. Cartularies*, 2 vols, R. Hist. S., Camd. 4th ser. xxxi, xxxiii, 1986–7, ii, 250; F. W. Maitland, *D. B. and Beyond*, Cambridge, 1897, pp. 373, 384; Hallam, *Set. and Soc.*, 9, 10, 13, 29, 31, 38, 53, 159–61; Wilton Ho. Survs Mans 1631, vol. ii, Wylye fos. 3(2)–6(5); PRO, D.L., Sp. Commn 1260; Leics. RO, Arch. Wills: Nic. English 1578 Barton Grange, Prestwold; as n. 88.

128 Vinogradoff, *Villainage*, 234; *Growth Manor*, 178.

129 Ibid. 178–9; A. W. Richeson, *Engl. Ld Measuring to 1800: instrs and practices* Soc. Hist. Technol. and M.I.T., Cambridge, Mass. and London, 1966, pp. 5, 9, 11, 13, 17, 19, 24, 29, 34, 36, 56–7, 65, 67, 69, 70, 77, 80, 82, 88, 99, 100, 106; Bateson et al., op. cit., ii, 367–9.

130 B. M. S. Campbell, 'Regional Uniqueness Engl. Fld Systs', *Ag. Hist. Rev.* 1981 xxix, 18, 19; 'Pop. Change', 176–7, 179, 185–7.

131 BL, Add. MS 34008 fo. 1; Add. R. 37534, 37554; Spufford, 'Rural Cams.', 95sqq.

132 James and Malcolm, *Bucks.*, 29; Col. Saltmarsh, 'Some Howdens. Vils', *Trans. E. R. Archaeol. Soc.* 1907 (1906–7) xiii, pt ii, 173; as n. 547; cf. Marshall, *E. Dept*, 238.

133 My *Agrarian Probs in 16th cent. and aft.*, 1969, pp. 99–102, 112–13; Stenton, *Docs Danelaw*, 140–1; J. Smith (Smyth), *Lives Berkeleys* ed. J. Maclean, 2 vols, 1883, i, 141, 160–1; G. Owen, *Descron Penbroks* ed. H. Owen, 4 pts, Cymmrodorion Rec. Ser. 1, 1892–1936, pt i, 61.

134 Straton, op. cit., 52, 242–3; A. L. Humphreys, *Mtls f. Hist. ... Wellington*, 4 pts, 1908–14, ii, 171; A. C. Chibnall, op. cit., 92, 273–4, 276; PRO, Exch. A.O., M.B. 390 fo. 37v; Parl. Survs, Lincs. 22 fos 1sqq.

135 Ault, *Fmg*, 19, 81–2, 148; *Husb.*, 55; my *Ag. Rev.*, 172.

136 Ibid., 156, 163.

137 Ibid., 54; *V. C. H. Sy*, iv, 430, 438.

138 Ibid., 438; Baker, 'Flds and Fms', 168–9; 'Fld Systs Holmesdale', 4, 7, 9, 10, 13, 22; H. W. Knocker, 'Evol. Holmesdale: no 3, Man. Sundrish', *Arch. Cant.* 1932 xliv, 209–10.

139 Marshall, *S. Cos*, ii, 395, 403; PRO, Exch. K.R., M.B. 40 fos 20–1.

140 H. B. Muhlfeld, *Surv. Man. Wye*, New York, 1933, pp. xxiii, xxvii–viii, xxxiii, xxxv–vi, lxxv–vi, 1sqq.; *V. C. H. Kent*, iii, 321.

141 Marshall, *S. Cos*, ii, 395, 403; A. R. H. Baker, 'Fld Systs Kent', ts. thes. Ph.D., London Univ. 1963, pp. 129–30; H. C. M. Lambert, *Hist. Banstead in Sy*, 2 vols, Oxford, 1912, i, 184–5, 199sqq.; *V. C. H. Sy*, iv, 430.

142 T. A. M. Bishop, 'Rot. Crops at Westerham', *Econ. Hist. Rev.* 1938–9 ix, 39–42; my *Ag. Rev.*, 54–5; Baker, 'Fld Systs Holmesdale', 15.

143 Marshall, *S. Cos*, ii, 394–5, 403.

144 Ibid., 394–5; *Sy A. C.* 1933 xli, 42–3; Sy RO, 34/3 surv. Gt Bookham 1614 fos 3, 14, 15; Lambert, op. cit., i, 187–8.

145 Owen, op. cit., i, 61; Baker, 'Fld Systs Holmesdale', 19; D. Roden, 'Frag. Fms and Flds in Chiltern Hills: 13th cent. and later', *Med. Studs* 1969 xxxi, 225, 231–3, 235; Homans, op. cit., 112–13; my *Agrarian Probs*, 34–5, 37–8; W. Davies, *Gen. View Ag. and Dom. Econ. S. W.*, 2 vols, 1814, i, 222; H. L. Gray, op. cit., 186sqq.

146 G. V. Jacks, *Soil*, 1954, pp. 11, 12, 192, 195–6, 198.

147 J. Blum, *Ld and Peasant in Ru. fr. 9th to 19th cent.*, New York, 1968, pp. 25–6, 515; M. Bloch in *Cam. Econ. Hist. Eur.*, i, 268–9.

148 Stenton, *A.-S. Engld*, 314, 476–7, 479–80, 515; Coulton, op. cit., 10; Whitelock, *A.-S. Wills*, 2, 4, 12, 16, 20, 24, 36, 54, 56, 68, 70, 74, 78, 80, 84, 86, 88, 90, 92, 94, 101, 112, 163, 165, 175; *Begs Engl. Soc.*, Harmondsworth, 1952, pp. 18, 82, 86, 94, 96, 101, 106, 108sqq.; E. John, *Ld Tenure in Early Engld*, Leicester 1960, pp. 14, 15, 17, 18; H. P. R. Finberg, *Lucerna*, 1964, pp. 146–7; *Early Chs Wessex*, Leicester, 1964, p. 147; in Finberg and Thirsk, op. cit., i, pt ii, 497–8, 507–10; A. L. Poole, *Fr. Dom. to Magna Carta*, Oxford, 1955, p. 40; *Obs Soc. in 12th and 13th cents*, Oxford, 1946, p. 11; Madox, op. cit., 416; F. E. Harmer, *Sel. Engl. Hist. Docs 9th and 10th cents.*, Cambridge, 1914, pp. 32–3, 116; T. H. Aston, 'Origs Man. in Engld', *Trans. R. Hist. S.* 5th ser. 1958 viii, 65, 69, 71–2.

149 Ibid., 65–6, 69, 71–3; Stenton, *A.-S. Engld*, 479–80; Finberg, *Lucerna*, 146–7; in Finberg and Thirsk, op. cit., i, pt ii, 510sqq.; Whitelock, *Begs*, 112.

150 BL, Add. MS 15350 fos 47 (49), 105 (106); Harl. MSS 436 fo. 65; 4660 fo. 9 (13); Cott. MSS Aug.ii.48; Claud.B.vi, fos 40, 61, 74v, 80v, 85v, 86v,

93v, 100v–1 (101v–2); C.ix, fos 116v, 121, 127 (125) (23); Tib.A.xiii, fos 66, 73v, 84v, 94, 96v; Bede, op. cit. 779; cf. Birch, op. cit., 217, 227sqq., 347–8.

151 Cf. V. Skipp, 'Evol. Set. and Open Fld Topog. in N. Arden dn to 1300' in Rowley, op. cit., 174.

152 A. C. Chibnall, op. cit., 276; Douglas, op. cit., 19; Baker, 'Flds and Fms', 166–7; 'Fld Systs Holmesdale', 9; Kerridge, *Survs*, 12, 14, 15, 18, 21sqq., 36, 39, 65, 79, 80, 104, 131, 136–7; Wilts. RO, Acc. 283 Amesbury Surv. Bk 1574, Billet ld.

153 Kemp, op. cit., ii, 282, 286–7; Vinogradoff, *Villainage* 233–5, 457–8; Stenton, *Docs Danelaw*, 202.

154 T. M. Charles-Edwards, 'Kinship, Status and Origs Hide', *Past and Pres.* 1972 no. 56, pp. 5–7, 14, 21; B. S. Phillpotts, *Kindred and Clan in M. A. and Aft.*, Cambridge, 1913, pp. 205sqq., esp. 216–18, 224, 228–9, 231sqq., 240sqq., 246, 251, 253–5; Whitelock, *Begs*, 37sqq.; Stenton, *A.-S. Engld*, 314sqq., 470sqq.; M. Longfield in J. W. Probyn, *Systs Ld Tenures in Var. Countries*, [1881] p. 27; J. E. A. Jolliffe, *Pre-Feudal Engld: Jutes*, 1933, pp. 24, 27; Dodgshon, *Origs*, 125; G. R. J. Jones in A. R. H. Baker and R. A. Butlin, *Studs Fld Systs in Brit. Is.*, Cambridge, 1973, pp. 432–4, 441–2, 446, 452; in Finberg and Thirsk, op. cit., i, pt ii, 320sqq.; 'Customary Tenures in W. and Open-Fld Ag.' in Rowley, op. cit., 203–6; 'Med. Open-Flds and ass. set. patts in N. W. W.' in *Géog. et Hist. Agraires*, Ann. de l'Est Mém. 21, Nancy, 1959, pp. 313sqq.; T. J. Pierce, 'Agrarian Asps Tribal Syst. in Med. W.', ibid., 329sqq.; J. E. Handley, *Scot. Fmg in 18th cent.*, 1953, pp. 92–3, 234–5.

155 My *Ag. Rev.*, 155–7; Sylvester, op. cit., 220–1, 236–9, 245–8, 331, 339–40, 446–8, 450–1, 471–3, 477, 484.

156 Pliny, *Naturalis Historiae* lib. xviii, cap. xlviii.

157 *Archiv f. lateinische Lexicographie* 1886 iii, 285.

158 Cf. W. H. Manning, 'Piercebridge Plough Gp' in G. de G. Sieveking, *Prehistoric and Rom. Studs*, 1971, pp. 132–4, pls XLI–VII.

159 A. G. Haudricourt and M. J.-B. Delamarre, *L'Homme et la charrue à travers le monde*, Paris, 1955, pp. 204sqq.

160 Virgil, *Georgics* bk i, ll. 19, 45–7, 67, 97, 162, 169–74, 261–2, 494; bk ii, l. 423; R. G. Collingwood and I. Richmond, *Archaeol. Rom. Brit.*, 1969, p. 311, pl. XX; W. H. Manning, 'Plough in Rom. Brit.', *Jnl Rom. Brit.* 1964 liv, 56sqq., 62–3, pl. VIII; pace K. D. White, *Ag. Impls Rom. Wld*, Cambridge, 1967, pp. 134, 140.

161 Haudricourt and Delamarre, op. cit., 16, 139.

162 J. B. Passmore, *Engl. Plough*, 1930, pp. 2, 7; Manning, 'Plough', 62–3; C. Hawkes, 'Rom. Villa and Hvy Plough', *Antiquity* 1935 ix, 340.

163 Haudricourt and Delamarre, op. cit., 352; Bowen, op. cit., 36; cf. E. Barger, 'Pres. Posn Studs in Engl. Fld Systs', *Engl. Hist. Rev.* 1938 liii, 394–5, 403.

164 F. G. Payne, 'Plough in Anc. Brit.', *Archaeol. Jnl* 1948 (1947) civ, 91–2, 97, 108–9; 'Brit. Plough: some stages in dev.', *Ag. Hist. Rev.* 1957 v, 78–9; Hawkes, art. cit., 340; Manning, 'Plough', 59, 65.

165 Haudricourt and Delamarre, op. cit., 109–10.

166 Ibid., 352–3.

167 E.C. Curwen, *Plough and Pasture* 1946, pp. 61–4; 'Anc. Cults at Grassington, Yks.', *Antiquity* 1928, ii, 168sqq.; Raistrick, op. cit., 57, 62; Bowen, op. cit., frontis., pp. 2, 23, 25, pls I–III.

168 W.G. Clarke, *In Breckland Wilds*, 1926, p. 22; L. Laing, *Celt. Brit.*, 1979, p. 38; J. Porter, 'Waste Ld Reclam. in 16th and 17th cents.; ca. S.E. Bowland 1550–1630', *Trans. Hist. Soc. Lancs. and Ches.* 1978 (1977) cxxvii, 19; my *Ag. Rev.*, 19, 77.

169 Ibid., 34, 50, 357; Bowen, op. cit., 21, pl. V; Laing, op. cit., 38.

170 Barger, art. cit., 391sqq., 403; Bowen, op. cit., 23–5, 36; W. Müller-Wille, 'Langstreifenflur u. Drubbel: ein Beitr. z. Siedlungs-geogr. Westgermaniens', *Dtsch. Archiv f. Landes- u. Volksforschung* 1944 viii, 30; my *Ag. Rev.*, 34–5, 67; E. Jutikkala, 'How Open Flds came to be divided into numerous Selions', *Proc. Fin. Acad. Sci. and Ltrs 1952*, Helsinki, 1953, pp. 119, 121, 125, 127, 132, 140.

171 D. Defoe, *Tour thr. E. W.*, 2 vols, E.E. i, 198.

172 Payne, 'Brit. Plough', 78; cf. S.C. Stanford, *Archaeol. M.A.*, 1980, pp. 79, 81–2, 113; Laing, op. cit., 52; W. Gardner and H.N. Savory, *Dinorben: hill-ft occ. in early iron age and Rom. times*, Cardiff, 1971, pp. 5, 6, 158.

173 P. Salway, *Rom. Brit.*, Oxford, 1981, pp. 621–2; Laing, op. cit., 38.

174 V. Chapman, 'Open Flds in W. Ches.', *Trans. Hist. Soc. Lancs. and Ches.* 1953 (1952) civ, 59; Sylvester, op. cit., 480–3 et pass.; H.L. Gray, op. cit., 242sqq.

175 Ibid., 226, 232, 271; my *Ag. Rev.*, 129, 144.

176 Ibid., 47–8, 57, 61, 72, 76–7, 79, 81, 105–7, 159, 161–2, 167; P.F. Brandon, 'Arable Fmg in Suss. Scarp-ft Par. during later M.A.', *Suss. A.C.* 1962 c, 67.

177 A. Steensberg, 'Sula: anc. term f. wheel plough in N. Eur.', *Tools and Tillage*, 1976–9 iii, 94–5; *O.E.D.*; my *Ag. Rev.*, 50, 357; J.E. Jackson, op. cit., 28, 164.

178 D.M. Wilson, 'A.-S. Rural Econ.', *Ag. Hist. Rev.* 1962 x, 75–6.

179 Ibid.; Haudricourt and Delamarre, op. cit., 358–9; P. McGurk, D.M. Dumville, M.R. Godden and A. Knock, *11th cent. A.-S. Illus. Misc.*, Copenhagen, 1983, p. 42; BL., Cott. MS Jul. A. vi, fo. 3(2); Tib.B.v pt i, fo. 3.

180 *O.E.D.*; Steensberg, art. cit., 91, 94, 96.

181 Ibid., 95–6; K.I. Sandred, 'N. on Engl. Plough', ibid., 259.

182 Müller-Wille, 'Langstreifenflur', 29, 30.

183 H.L. Gray, op. cit., 350sqq., esp. 354; Douglas, op. cit., 26; J.Z. Titow, *Engl. Rural Soc. 1200–1350*, 1969, pp. 162–4; R.H.C. Davis, 'E.A. and Danelaw', *Trans. R. Hist. Soc.* 5th ser. 1955 v, 27.

184 Stenton, *A.-S. Engld*, 278sqq.; J.N.L. Myres, *Engl. Sets*, Oxford, 1986, pp. 79–81, 85, 128, 142–3, 151sqq.; R.R. Clarke, *E. A.*, 1960, pp. 129–30, 132–3.

185 Titow, *Engl. Rural Soc.*, 162–4; Douglas, op. cit., 17, 37; Campbell, 'Pop. Chg.', 178; H.L. Gray, op. cit., 93sqq., 303; Baker, 'Flds and Fms', 164–5; Jolliffe, op. cit., 24; *V.C.H. Sfk*, i, 641–4; *Shotley Par. Recs.* Sfk Grn Bks xvi, pt 2, Bury St Edmunds, 1912, pp. 266sqq.; PRO, D.L., R. and S. 9/13 mm. 3sqq.; Exch. L.R., M.B. 201 fos 217sqq.; 220 fos 227, 229–30; BL, Add. Ch. 10229; Add. MS 2349 fos 1sqq.

186 H. L. Gray, op. cit., 246sqq.; A. N. Palmer, *Tn, Flds and Folk Wrexham in time Jas I*, Wrexham and Manchester, n.d., frontis., p. 20; D. G. Hey, *Engl. Rural Commun.: Myddle under Tudors and Stuarts*, Leicester, 1974, pp. 29–31; Sylvester, op. cit., 220, 229–30, 241sqq., 273, 277–8, 287sqq., 304, 307–9, 311–12, 316, 335–6, 340, 451sqq., 459, 476, 479, 482–3, 487–90, 493–5; 'Set. Patts in Rural Flints.', *Flints. Hist. Soc. Pubs* 1954–5 xv, 19sqq., 25–6, figs 2, 5–8; G. R. J. Jones in Finberg and Thirsk, op. cit., i, pt ii, 344sqq.; E. Miller, ibid. vol. ii, 400; R. I. Jack, ibid., 412–13; G. White, 'Glimpses of Open-Fld Landscape', *Ches. Hist.*, 1988 no. 21, pp. 8, 9, 11, 12, 16; P. Dodd, 'Landscape Hist.', ibid., 25–6; M. C. Hill, 'Wealdmoors 1560–1660', *Trans. Shrops. Archaeol. Soc.* 1953 (1951–3) liv, 298sqq.; M. F. Howson, 'Aughton, nr Lanc.', *Trans. Lancs. and Ches. Ant. Soc.* 1959 lxix, figs III–V; my *Ag. Rev.*, 129, 145, 157.

187 Ibid., 53–4; Muhlfeld, op. cit., pp. lvi, lxxvi; N. Neilson, *Cartulary and Terrier Priory Bilsington, Kent*, 1928, pp. 13sqq., 39sqq., 148sqq.; Baker, 'Open Flds and Partible Inheritance', 21–2; Jolliffe, op. cit., 1sqq., 24, 27, 93sqq., 116sqq.

188 Ibid., 22–4; Homans, op. cit., 110–12.

189 Jolliffe, op. cit., 19sqq.; R. H. D'Elboux, *Survs Mans Robertsbridge, Suss. and Michelmarsh, Hants. and Demesne Lds Halden in Rolvenden, Kent, 1567–70* Suss. Rec. Soc. 1944 xlvii, 2, 3; Straton, op. cit., 1sqq.; Neilson, *Terrier Fleet* 7sqq.; Lancs. RO, DD.F 112; Glouc. Ref. Lib. MS 16062 pp. 25, 53; Bucks. RO, 155/21 ct files Taplow 2 May 1620; Northants. RO, Mont. Coll. Northants. old box 17 no. 160 Ct Bk Broughton, Ct Recog. 3 June 1 Ed. 6; Wilton Ho., Survs Mans in Co. Wilts. fos 1sqq.; PRO, Exch. A. O., Parl. Survs: Corn. 9 fos 2, 3; 23 fos 7–10; 38 fo. 2; Devon 14 fo. 1.

190 Muhlfeld, op. cit., xxxiii, lxxv; Baker, 'Flds and Fms', 157, 159sqq.; 'Fld Systs Holmesdale', 4, 7, 9, 13, 19, 22; Jolliffe, op. cit., 13; A. Everitt, *Continuity and Colonization: evol. Kentish set.*, Leicester, 1986, pp. 69, 70, 76; my *Ag. Rev.*, 54; pace H. L. Gray, op. cit., 282sqq.

191 Ibid., 410sqq.

192 My *Ag. Rev.*, 154–5, 157; Sylvester, op. cit., 112, 121, 220–1, 480–3; P. Vinogradoff and F. Morgan, *Surv. Hon. Denbigh 1334* 1914, pp. xlv sqq., liii, 1, 2, 4, 6, 9sqq., 18, 20sqq., 52, 56–7, 62sqq., 83–4, 86sqq., 110–13, 115–16, 121–3, 131sqq., 137–9, 147, 204, 230–2, 252–5, 258.

193 H. L. Gray, op. cit., 71–2.

194 Ibid., 272sqq., 401; Sylvester, op. cit., frontis. et pass.

195 Vinogradoff, *Engl. Soc.*, 273–4; K. C. Newton, *Thaxted in 14th cent.*, Chelmsford, 1960, pp. 33–5.

196 Spufford, 'Rural Cambs.', 95sqq.

197 Cf. Fox, 'Approaches', 101.

198 Humphreys, op. cit., pt ii, 171, 232sqq.; Marshall, *W. Engld* ii, 169, 196; Soms. RO, DD/X/GB Combe Surv. Bk 1704 pp. 103sqq.; DD/SP Acc. H/62 box 58, Surv. Taunton Cas. and Deane 28 Apr. 8 Eliz.; Taunton Deane Enclo. Award 1851; my *Ag. Rev.*, 115, 132, 134, 170.

199 Ibid., 54, 78, 84–5, 87, 89, 136; W. Hudson, 'Traces Prim. Ag. Org. as suggested by Surv. Man. Martham, Nfk (1101–1292)', *Trans. R. Hist. S.* 4th ser. 1918 i, 38–9; Kemp, op. cit., i, 340; Douglas, op. cit., 18sqq., 238–9, 241, 243; K. M. Dodd, op. cit., 88, 152, 157.

200 Ibid., 88, 152, 155, 157; Burrell, op. cit., 25, 28–30, 33, 46, 108; F. Hull, 'Ag. and Rural Soc. in Ess. 1560–1640', ts. thes. Ph.D. London Univ. 1950, pp. 12, 13, 25–6; N. Evans, 'Commun. S. Elmham, Sfk, 1550–1640', ts. thes. M.Phil. Univ. East Anglia 1978, p. 67; Sfk RO (Ips.) V.5.22.1 fos 5v, 6, 11v, 12, 13v, 14, 17v, 18, 23v–4, 27v–8; 50.1.74 (1), (12); BL, Add. MSS 21054 fos 2sqq.; 23950 fo. 98; 32134 fos 11 (3) (4) sqq., 29 (22) pp. 1sqq.; as nn. 137–44, 184–5, 189–90.

201 B. M. S. Campbell, 'Ag. Prog. in Med. Engld: some evid. fr. E. Nfk', *Econ. Hist. Rev.* 2nd ser. 1983 xxxvi, 39 et pass.; 'Pop. Change', 174, 176–9, 181; 'Regional Uniqueness', 20–2; Millican, op. cit., 83sqq.; J. A. Venn, 'Econ. Nfk Par. in 1783 and at Pres. Time', *Econ. Hist.* 1929 i, 77; my *Ag. Rev.*, 87–8.

202 Ibid., 118, 120–3, 126; H. S. A. Fox, 'Fld Systs E. and S. Devon: Pt i, E. Devon', *Devon Assoc.* 1972 iv, 95–7, 101sqq., 107, 113–15, 117; Smith, *Lives Berkeleys*, i, 141, 160–1; E. S. Lindley, 'Hist. Wortley in Par. Wotton-under-Edge', *T. G. B. A. S.* lxviii–ix 1949–50, lxix, 168, 178; PRO, Exch. L.R., M.B. 207 fos 20 (18)v, 22 (20); Sylvester, op. cit., 220–1, 239, 396, 398sqq.

203 Ibid., 220–1, 245–8, 446–8, 450–1, 471–3, 477, 484; my *Ag. Rev.*, 150–1, 157, 159, 162, 166, 169, 172; Jones in Finberg and Thirsk, op. cit., i, pt ii, 340–1, 351–3; W. A. Hulton, *Coucher Bk or Cartulary Whalley Abb.*, 3 vols, Chetham Soc. x, xi, xvi, 1847–8, iii, 647–8; A. J. Petford, 'Process Enclo. in Saddleworth, 1625–1834', *Trans. Lancs. and Ches. Ant. Soc.* 1987 lxxxiv, 82–3; Darlington, *Darley Abb.* 389–90, 400–1; W. E. Wightman, 'Open-Fld Ag. in Peak Dist.', *Derbys. Archaeol. Jnl* 1961 lxxxi, 115sqq., 123; J. C. Jackson, 'Open-Fld Cult. in Derbys.', ibid. 1962 lxxxii, 64; J. P. Carr, 'Open-Fld Ag. in Mid-Derbys.', ibid. 1963 lxxxiii, 69sqq.

204 W. Marshall, *On Appropriation and Inclosure Commonable and Intermixed Lds*, 1801, p. 10.

205 My *Ag. Rev.*, 172; R. S. Dilley, 'Cumb. Ct Leet and Use Com. Lds', *Trans. Cumb. and Wmld Ant. and Archaeol. Soc.* n.s. 1967 lxvii, 133–5; T. H. B. Graham, *Bny Gilsland: Ld Wm Howard's Surv. taken in 1603* Cumb. and Wmld Ant. and Archaeol. Soc. extra ser. 1934 xvi, 60–1, 72sqq., 102sqq., 126–7, 133sqq.; A. P. Appleby, *Famine in Tu. and Stuart Engld*, Liverpool, 1978, pp. 41–3; G. G. Elliott, 'Syst. Cult. and Evid. Enclo. in Cumb. Open Flds in 16th cent.' in *Géog. et Hist. Agraires* Ann. de l'Est Mém. 21, pp. 119, 122–4.

206 Raistrick, op. cit., 17, 42, 54–5, 68; Lancs. RO, DD.X,160/3; PRO, Exch. L. R., M.B. 193 mm. 104 (36) sqq., 112 (44); 195 mm. 1sqq., 38–41, 64–6, 67d, 72–4, 75dsqq., 95–6; Bateson et al., op. cit., iii, 102; iv, 317, 328, 357, 381, 403; x, 133–6, 270, 276, 366–7, 387; xii, 142, 157, 184, 191, 239, 318; Dilley, art. cit., 133–5; Hartley and Ingilby, op. cit., 7, 238.

207 Ibid., 241–2; Raistrick, op. cit., 43; Bateson et al., op. cit., iv, 317; x, 270, 276, 366–7; xii, 142, 158–9, 165, 239, 318–19.

208 Ibid., x, 387–8.

209 Ibid., xii, 157, 166; J. P. Rylands, *Lancs. Inqs retd into Chanc. D. L., Stuart Per.* 3 pts, Rec. Soc. f. Lancs. and Ches. iii, xvi, xvii 1880–8, i, 257; ii, 287; PRO, Exch. L.R., M.B. 195 mm. 64, 67d–8, 90sqq.; C. M. L. Bouch and G. R. Jones, *Sh. Econ. and Soc. Hist. Lake Cos 1500–1830*, Manchester, 1961, p. 77; Raistrick, op. cit., 20.

210 My *Ag. Rev.*, 162; E. J. Evans and J. V. Beckett in Finberg and Thirsk, op. cit., v, pt i, 19, 20; Raine, *Priory Hexham*, ii, 50.

211 Marshall, *W. Engld*, ii, 112.

212 Ibid., ii, 196; H. S. A. Fox, 'Chron. Enclo. and Econ. Dev. in Med. Devon', *Econ. Hist. Rev.* 2nd ser. 1975 xxviii, 183–4, 187–90, 'Fld Systs ... E. Devon', 83–4; A. H. Shorter, 'Fld Patts in Brixham Par.', *Devon Assoc.* 1950 lxxxii, 273sqq.; A. H. Slee, 'Open Flds Braunton', ibid., 1952 lxxxiv, 142sqq.; Venn, *Founds*, 45; J. Hatcher, *Rural Econ. and Soc. in Du. Corn. 1300–1500*, Cambridge, 1970, pp. 10, 11; in Finberg and Thirsk, op. cit., ii, 385; BL, Egerton MS 3134 fos 217sqq.; H. P. R. Finberg, *Tavistock Abb.*, Cambridge, 1951, pp. 46sqq.; 'Open Fld in Devon', *Antiquity* 1949 xxiii, 180sqq.; W. G. Hoskins and H. P. R. Finberg, *Devon Studs* 1952, pp. 265sqq.; my *Ag. Rev.*, 151.

213 PRO, Exch. L.R., M.B. 207 fo. 42 (40)v; cf. R. Carew, *Surv. Corn.*, (1602) 1769, p. 38 recte (30 erron.).

214 Marshall, *W. Engld*, i, 32, 100; ii, 45–7, 53, 131, 134–5, 172; *E. Dept*, 102; *Appropriation*, 10; Hatcher, op. cit., 10, 14; H. A. Lomas, *Hist. Abbotsham*, p.p., 1956, p. 7; Finberg and Thirsk, op. cit., i, pt ii, 494; Fox, 'Chron.', 187; BL, Harl. MS 5827 fo. 8 (6)v; Hoskins and Finberg, op. cit., 330–1.

215 Ibid., 284–6; *Cals Procs in Chanc. in reign Q. Eliz.*, ii, 248; iii, 264; BL, Add. R. 13858, 23 June 17 Jas; Harl. MS 71 fo. 56 (67); PRO, Exch. A.O., Parl. Survs, Corn. 9 fo. 16.

216 Ibid., A.O., M.B. 388 fos 143, 145, 147, 149, 151, 153, 155, 159, 167 (pp. 3sqq., 11, 15); A. L. Rowse, *Tu. Corn.*, 1941, pp. 33–5; J. Rowe, *Corn. in Age Ind. Rev.*, Liverpool, 1953, pp. 213–14; Hoskins and Finberg, op. cit., 284–6, 330–1; Soms. RO, DD/HP, Acc. C. 39 box 34 transcript arts agrt W. Buckland, lds fms and tens 1634 art. 4.

217 My *Ag. Rev.*, 159; Butlin in Baker and Butlin, op. cit., 98sqq., 133–4; E. Hughes, *N. Country Life in 18th cent.: N. E. 1700–50*, 1952, pp. 135–6.

218 Bateson et al., op. cit., ix, 324.

219 Ibid., ii, 369, 381–2, 418, 422, 478; v, 201, 424; xiv, 212; J. U. Nef, *Rise Brit. Coal Ind.*, 2 vols, 1932, facing p. 307.

220 Bateson et al., op. cit., i, 274; v, 488; *Wills and Invs illus. hist., manners, lang., stats, etc. N. Cos Engld fr. 11th cent. dnwds*, pt i, Surtees Soc. 1835, pp. 335sqq.; J. S. Purvis, *Sel. 16th cent. Causes in Tithe fr. York Dioc. Reg.* Yks. Archaeol. Soc. Rec. Ser. 1949 (1947) cxiv, 100.

221 PRO, Exch. L.R., M.B. 192 m. 32 (29); Bateson et al., op. cit., i, 234–5, 287; v, 202–3, 376–7, 416–17, 421–2; ix, 119.

222 Ibid., i, 351, 355; ii, 128–9, 367sqq, 416, 418–19, 458, 478; v, 202, 332, 372–3, 424, 488, 497; vii, 305, 307, 315; ix, 325–6; xii, 184; xiv, 211–13; G. R. Batho, *Hsehld Pprs Hy Percy 9th E. Nthld (1564–1632)* R. Hist. S. Camd. 3rd ser. 1962 xciii, p. li.

223 H. L. Gray, op. cit., 192–3, 202; Dodgshon, *Origs*, 32–3, 38, 40; 'Interpn Subdiv. Flds: stud. in pop. commun. ints', in Rowley, op. cit., 133–4, 141, 144–5; Jones, ibid. 203, 205–6, 209–10, 213–15; in Baker and Butlin, op. cit., 431sqq., 446sqq.; R. H. Buchanan, 'Fld Systs Ire.', ibid., 580sqq.; G. Whittington, 'Fld Systs Scotld', ibid., 530sqq.; M. Davies, 'Fld Systs S. W.', ibid., 480–3, 506sqq.; W. Greig, *Gen. Rep. on Gosford Estes in Co. Arm.*

1821 ed. F. M. L. Thompson and D. Tierney, Belfast, 1976, pp. 32, 104–5, 110, 171, 186; Owen, op. cit., pt i, 59sqq.; J. Otway-Ruthven, 'Org. Anglo-Ir. Ag. in M. A.', *Jnl R. Soc. Antiquaries Ire.* 1951 lxxxi, 2sqq.; Handley, op. cit., 38sqq.; W. Marshall, *Gen. View Ag. Cen. Highlands Scotld*, 1794, pp. 29, 30, 39; M. Gray, 'Regs and their Issues: Scotld' in G. E. Mingay, *Victorian Countryside*, 2 vols, London and Boston, 1981, i, 91; McCourt, art. cit., 369sqq.; my *Ag. Rev.*, 155, 157.

224 Ibid., 57; A. H. Cooke, *Early Hist. Mapledurham*, Oxford, 1925, p. 198; as n. 176.

225 D. Woodward, op. cit., 59; my *Ag. Rev.*, 61.

226 Ibid., 79.

227 Ibid., 105–6; Homans, op. cit., 84; F. M. Eden, *State Poor*, 3 vols, 1797, iii, 749.

228 PRO, Exch. L.R., M.B. 207 fo. 28 (26)v.

229 Hoare, op. cit., Hund. Alderbury, 69sqq.; Hund. S. Domerham, 43sqq.; Straton, op. cit., 131sqq., 163sqq.

230 E.g., C. S. Ruddle, 'N. on Durrington', *W. A. M.* 1903–4 xxxiii, 271; Slater, *Engl. Peasantry*, 20; Straton, op. cit., 6sqq., 35sqq., 52sqq., 96sqq., 111, 140sqq., 160–3, 232–5, 251sqq., 277sqq.

231 Ibid., 72sqq., 152–5, 207sqq.; PRO, Exch. L.R., M.B. 191 fos 126sqq.

232 Straton, op. cit., 124–6, 167sqq., 269sqq., 290sqq.; E. J. Bodington, art. cit., xli, 30, 117–18; Hoare, op. cit., Hund. Chalk 153–4; Wilts. RO, Savernake Coll. survs: Chisbury 1552, Burbages 1574, Collingbourne Kingston 1552, 1595, 1639, Collingbourne Ducis 1608, 1635, Shalbourne Westcourt 1639; Ct Bk Collingbourne Ducis temp. Eliz.; Chisbury custumal 1639; Acc. 283 Liber Supervisus Amisburie; Amesbury Surv. Bk; Surv. Amesbury 1635; Surv. Urchfont *c.* 1640; Bk exts Ct R. Var. Seymour mans; Wilton Ho. Survs Mans 1631 vol. i, Foffonnt; Bulbridge, Washerne and S. Ugford fo. 7 (8); vol. ii, Stanton Barnard fos 2sqq.; Survs Mans 1632–3 Chilmark and Rudge fos. 5–8, 10, 11; Stanton Ct Bk beg. 1724 ao 1729; PRO, D.L., M.B. 115 fo. 30; 117 fos 44sqq.; R.&S. 9/31; Sp. Commn 700; Exch., L.R., M.B. 187 fos 93, 98sqq., 135vsqq., 216vsqq., 361vsqq., 375vsqq., 395sqq.; 203 fos 365sqq.; C.P., Recovery R. 909, 50 Geo. 3 Trin. rots 160 (18) sqq.; Sp. Colls, Maps and Plans L/55, i, ii.

233 Baigent and Millard, op. cit., 324, 340, 348, 617, 623sqq.; Titow, op. cit., 145.

234 Jolliffe, op. cit., 11; G. M. Cooper, art. cit., 229, 240–1; J. C. K. Cornwall, 'Agrarian Hist. Suss. 1560–1640', ts. thes. M.A., London Univ. 1953, pp. 48–9; W. H. Godfrey, *Booke of Jn Rowe* Suss. Rec. Soc. 1928 xxxiv, 62–3, 67, 70, 107, 219, 246; Brandon, 'Arable Fmg', 61–2, 65.

235 Northants. RO, Wmld Coll. 5. ix, surv.

236 Ibid., survs Rottingdean, Rodmell; C. Thomas-Stanford, *Abs Ct R. Man. Preston (Preston Episcopi)* Suss. Rec. Soc. 1921 xxvii, 12, 25; BL, Add. MS 6027 fos 118 (117), 130 (120)v–131 (121); Godfrey, op. cit., 62–3, 107.

237 Marshall, *S. Dept* 8; H. L. Gray, op. cit., 371sqq.; Whiteman, op. cit., 99, 103, 108; Austin, op. cit., ii, 9, 10, 271–3; Seebohm, op. cit., frontis.; Roden in Baker and Butlin, op. cit., 329, 339; [Hervey], op. cit.; D. Cromarty,

Flds Saffron Walden in 1400, Chelmsford, 1966, pp. 8, 9; Walker, op. cit., 49; C. Vancouver, *Gen. View Ag. Co. Ess.*, 1795, p. 105; T. Batchelor, *Gen. View Ag. Co. Bedford*, 1808, p. 538; James and Malcolm, *Bucks.*, 28; G. Longman, *Cnr Engld's Gnd*, 2 vols, s.l. n.d., ii, 5, 19; D. Bushby and W. Le Hardy, *Wormley in Herts.*, 1954, p. 67; Cooke, op. cit., 114, 195–8; Hull, op. cit., 15–18, 21–3, 26–8; J. B[lagrave], *Epit. Art Husb.*, 1669, p. 219; J. Spratt, 'Agrarian Conds in Nfk and Sfk, 1600–50', ts. thes. M.A., London Univ. 1935, pp. 236–7; M. Spufford, *Contrasting Communs: Engl. Villagers in 16th and 17th cents*, Cambridge, 1974, p. 60; cf. J. G. Jenkins, *Cartulary Missenden Abb.* pt i, Bucks. Archaeol. Soc. Recs Brch, 1938, ii, 66, 68–9, 184–5.

238 Ibid., 62; Cooke, op. cit., 196; Newton, *Thaxted*, 33–5; Homans, op. cit., 66, 422–3; A. E. Levett, *Studs in Man. Hist.*, Oxford, 1938, p. 183; Bucks. RO, P. 24/3 Ct R. Pitstone 3 Oct. 1665; Herts. RO, Cashiobury and Gape Coll. 8474; Moulton Coll. 44467 Ct R. Pirton 1696; G'bury Coll. xi.11 fos 203–5; Misc. MS 41332 Ct Bk Hexton 1647; Marshall, *Midl. Dept*, 561; Seebohm, op. cit., 11.

239 Hallam, *Set. and Soc.*, 230sqq.; A. Harris, *Open Flds E. Yks.* E. Yks. Loc. Hist. Soc. 1974, pp. 4, 6, 7; my *Ag. Rev.*, 60–2.

240 Ibid., 63.

241 R. V. Lennard, 'Engl. Ag. under Chas II: evid. R. S.'s "Enqs" ', *Econ. Hist. Rev.* 1932–4 iv, 42; Straton, op. cit., 409sqq.; W. Marshall, *Rev. Reps to Bd Ag. fr. W. Dept Engld*, 1810, p. 406; R. H. Gretton, *Burford Recs*, Oxford 1920, pp. 156, 353–5, 625–7, 673, 676, 679; C. C. Brookes, *Hist. Steeple Aston and Mid. Aston*, Shipston-on-Stour, 1929, p. 171; A. Ballard, 'N. on Open Flds Oxon', Oxf. Archaeol. Soc. *Rep.* 1908 (1909) pp. 24–7; 'Tackley in 16th and 17th cents', ibid. 1911 (1912) pp. 32–3; *W. A. M.* 1905–6 xxxiv, 296; Phillimore and Fry, op. cit., i, 70–1, 73sqq., 85.

242 My *Ag. Rev.*, 72.

243 H. L. Gray, op. cit., 313–5, 324; Spratt, op. cit., 36–7, 39, 51, 55–6; Corbett, art. cit., 70–1; R. H. Tawney, *Agrarian Prob. in 16th cent.*, 1912, pp. 254–5; BL, Stowe MS 765 fos 9sqq.; Add. R. 19082–4; PRO, Exch. L.R., M.B. 201 fos 65sqq.; Nfk RO, D.&C. MS, Parl. Surv. fos 98sqq.

244 Sfk RO (Ips.) V.11.2.1.1; Postgate, art. cit. 97; M. Bailey, *A Marg. Econ.? E. A. Breckland in later M. A.*, Cambridge, 1989, pp. 41, 57, 59.

245 Millican, op. cit., 83, 205sqq.

246 BL, Add. MSS 36745 fos 2sqq.; 40063 fos 70sqq.; Stowe MS 870 fos 3vsqq., 22v, 61, 83–5 (pp. 4sqq., 40, 117, 161, 163–6); PRO, Exch. L.R., M.B. 220 fos 190 (2) sqq.; Postgate in Baker and Butlin, op. cit., 298–9.

247 Colman, op. cit., 285.

248 PRO, Exch. L.R., M.B. 221 fos 8sqq., 52; Northants. RO, Misc. Led. 145 pp. 41, 371; Maps 1370, 1380.

249 W. Mavor, *Gen. View Ag. Berks.*, 1808, p. 166; BL, Add. MS 36903 fos 1sqq., 208v; Add. R. 49325.

250 Orwin, op. cit., 116.

251 Northants. RO, Wmld Coll. 4.xii.6; PRO, Exch. L.R., M.B. 221 fos 121–2; Deene Ho., Brudenell MSS E.vi.26a–b; I.v.22, 1 Oct. 9 Eliz.; O.xxiv.4 fo. 20.

252 Ibid. J.xx.9a, 11b; O.viii.17; J.C. Jackson, art. cit., 58; Seebohm, op. cit., 11; Roden in Baker and Butlin, op. cit., 352; PRO, D.L., M.B. 109 fos 61sqq., 110v–11; Exch. L.R., M.B. 201 fo. 18; 208 fos 138sqq.; 222 fos 155sqq., 179; 224 fos 17sqq.; Birm. Ref. Lib. MS 321480; Bucks. RO, D/MH 28/1; Northants. RO, Mont. Coll. box 25 no. 55; Misc. Led. 125; 145 pp. 70–1, 184; BL, Add. MS 18458 fos 1 (3) vsqq., 101; Beds RO, C. 364; PA.161; AD.1060.

253 Ibid., L.26/280–1.

254 M. Bateson, *Recs Boro. Leic.*, 3 vols, Cambridge, 1905 iii, 453–4; C.J. Billson, 'Open Flds Leic.', *T. L. A. S.* 1925–6 xiv, 2, 10, 16, 17; F.G. Emmison, *Types Open Fld Pars in Midlds* H.A. Pamph. 108, 1937, pp. 6, 7, 9; Leics. RO, Beaumanor Coll. Surv. Beaumanor 1608; Ct R. Barrow-on-Soar 1560.

255 J. Wake, 'Communitas Villae', *Engl. Hist. Rev.* 1922 xxxvii, 409–10; I. Beckwith, 'Remodelling Com. Fld Syst', *Ag. Hist. Rev.* 1967 xv, 109–10.

256 My *Ag. Rev.*, 142–4; J.R. Ravensdale, *Liable to Floods: vil. landscape on edge fens A.D. 450–1850*, Cambridge, 1974, pp. 85sqq., 90–2, 96, 99–101, 106–7.

257 Northants. RO, F (M) Coll. Misc. Vol. 99; Cambs. RO, L.1/20 fos 6, 7, 9; BL, Add. MS 33452 fos 33v–4, 37vsqq., 45sqq., 72v–3, 92vsqq., 99v, 123–4; M.P. Hogan, 'Clays, *Culturae*, and Cultivator's Wisd.', *Ag. Hist. Rev.* 1988 xxxvi, 125, 130.

258 H.L. Gray, op. cit., 37, 93sqq., 145sqq., 153, 521–3; Sylvester, op. cit., 220–1, 228–9, 231–4, 351, 359–60, 365sqq., 389–92, 405–6, 410, 429sqq.; J.A. Sheppard, *Origs and Evol. Fld and Set. Patts in Herefs. Man. Marden* Q.M.C., Geog. Dept Occ. Ppr xv, 1979, pp. 4–6, 8, 12; Glos. RO, D.326/E.1 fos 73sqq., 89v, 94v–5, 99v, 102v; PRO, Exch. L.R., M.B. 185 fos 86vsqq.; 217 fos 38sqq., 49sqq., 210sqq., 310sqq., 321sqq., 328–9, 331sqq., 342sqq., 362sqq.; M.B. 122; BL, Harl. MS 7639 (bd w. MS 70) fos 2v, 5, 6, 19, 23; my *Ag. Rev.*, 147.

259 Ibid., 114–15, 118, 123.

260 PRO, Exch. L.R., M.B. 207 fo. 20 (18)v; Glos. RO, D. 127/608.

261 Sylvester, op. cit., 220–1; my *Ag. Rev.*, 121–2.

262 Ibid., 92–4; Wake, art. cit., 410; A.C. Chibnall, op. cit., 227; Ault, *Husb.* 20, 32–3, 91; H.L. Gray, op. cit., 66, 452, 456–7, 460, 470–1, 476, 484, 491, 494, 497–8, 504, 508–9; *V.C.H. Leics.*, ii, 160–1; Spufford, *Contrasting Communs*, 96; A. Harris, *Open Flds E. Yks.* E. Yks. Loc. Hist. Soc. 1974, p. 6; Darlington, *Darley Abb.*, 102–3; Sheppard in Baker and Butlin, op. cit., 170; Postgate, ibid., 296–8; Roden, ibid., 352; F.M. Page, *Wellingborough Man. Accts* Northants. Rec. Soc. 1936 viii. p. xxiv; *Estes Crowland Abb.*, Cambridge, 1934, p. 119; S. Elliott, 'Open-Fld Syst Urban Commun.: Stamford in 19th cent.', *Ag. Hist. Rev.* 1972 xx, 156; Rec. Cmmn, *Placitorum in Domo Capitulari Westmonasteriensi Asservatorum Abbreviatio temporibus Ric. I, Johann., Henri. III, Ed. I, Ed. II*, 1811, p. 251; C. Dyer, *Lds and Peasants*, 322; Barratt, op. cit., ii, 180sqq.; Hallam in Finberg and Thirsk, op. cit., ii, 272; J. Raftis, ibid., 327–8; Dyer, ibid., 369–70; M.A. Havinden, 'Ag. Prog. and Open-Fld Oxon.' in E.L. Jones, *Ag. and Econ. Growth in Engld 1650–1815*, 1967, pp. 73–4; Oxon. RO, Wi.x/34, 2nd p. 98; Beds. RO, C.364; cf. P.D.A. Harvey, *Oxon. Med. Vil.: Cuxham, 1240–1400*, 1965, pp. 17, 18.

263 My *Ag. Rev.*, 114, 121; Dyer in Finberg and Thirsk, op. cit., ii, 369; Marshall, *Glos.* i, 64, 136; *W. Dept* 398; G. Turner, *Gen. View Ag. Co. Glouc.*, 1794, p. 35; PRO, Chanc. Proc. ser. i, Jas B.12/68 m. 3; Glos. RO, D.184/M11, p. 4; M. 24, p. 139; Homans, op. cit., 58; H. L. Gray, op. cit., 466, 496, 504.

264 My *Ag. Rev.*, 117; cf. H. L. Gray, op. cit., 509.

265 My *Ag. Rev.*, 126; W. H. Hart, *Historia et Cartularium Monasterii Sancti Petri Gloucestriae*, 3 vols, Rolls ser. 1863–7, i, 289sqq., 295; Hilton, *Med. Soc.*, 67–8.

266 My *Ag. Rev.*, 118; H. A. S. Fox, 'Alleged Transformation fr. 2-fld to 3-fld Systs in Med. Engld', *Econ. Hist. Rev.* 2nd ser. 1986 xxxix, 530; M. Whitfield, 'Med. Flds S. E. Soms.', *Soms. Archaeol. and Nat. Hist.* 1981 cxxv, 21, 24–5; cf. H. L. Gray, op. cit., 462, 496.

267 My *Ag. Rev.*, 123; cf. H. L. Gray, op. cit., 503.

268 My *Ag. Rev.*, 143; Vancouver, *Cambs.*, 129–31, 133, 147; Spufford, *Contrasting Communs* 120, 128–9; Hallam, *Set. and Soc.* 172–3; Page, *Crowland Abb.*, 119, 140, 444; Ravensdale, op. cit., 91–2, 106; M. W. Barley, 'E. Yks. Man. Bye-laws', *Y. A. J.* 1943 xxxv, 52; Hogan, art. cit., 125, 130; BL, Add. MS 33452 fos 37vsqq., 45sqq., 72v–3, 92vsqq., 99v, 123–4; Rec. Commn, *Rotuli Hundredorum temp. Hen. III et Edw. I in Turr. Lond. et in Curia Receptae Scaccarii West. asservati*, 2 vols, 1812–18, ii, 462, 605; cf. H. L. Gray, op. cit., 460, 476.

269 Sylvester, op. cit., 241, 273, 482; 'N. on Med. 3-cse Arable Systs in Ches.', *Trans. Hist. Soc. Lancs. and Ches.* 1959 (1958) cx, 183–4, 186; Jack in Finberg and Thirsk, op. cit., ii, 413–14; Jones in Baker and Butlin, op. cit., 465; Vinogradoff and Morgan, op. cit., xlviii–l, 1, 2, 4, 52, 56–7, 230; L'pool RO, Norris Pprs no. 552; Ches. RO, Cons. Ct Cestr. supra inv. Th. Ackerley 1675 Christleton; my *Ag. Rev.*, 129; cf. H. L. Gray, op. cit., 66, 498.

270 Owen, op. cit., pt i, 59, 60, 63.

271 My *Ag. Rev.*, 58; Rec. Cmmn, *Placitorum in Domo*, 306; Postgate in Baker and Butlin, op. cit., 298; Roden, ibid., 335; Newton, *Thaxted*, 33–5; Raftis in Finberg and Thirsk, op. cit., ii, 29–30; Holderness, ibid., v, pt i, 210; Homans, op. cit., 66, 422–3; Seebohm, op. cit., 11, 450–1; R. L. Hine, *Hist. Hitchin*, 2 vols, 1927, i, 65; Cromarty, op. cit., 5, 6; Levett, op. cit., 183, 338–9; Herts. RO, Arch. St Albans inv. Jn Phippe 1684 Norton; Moulton Coll. 44467 Ct R. Pirton 1696; Dimsdale Coll. 1 Ct R. Meesdenbury (Missenden) 7 Apr. 16 Jas; Misc. MSS 41673 Ct R. 14 May 3 Chas; 65810 pp. 143sqq.; cf. H. L. Gray, op. cit., 80, 490.

272 Northants. RO, Mont. Coll. Northants. old box 7 no. 66/5 Ct files Weekley 16 Apr. 25 Eliz.

273 BL, Add. MS 37682 fos 1sqq.

274 Morton, op. cit., 35.

275 H. Gill and E. L. Guilford, *Rect's Bk Clayworth 1672–1701* Nottm 1910, pp. 99, 103, 106, 111, 116, 121, 124, 127, 133, 141.

276 Marshall, *W. Dept* 398; *Glos.*, i, 65, 68, 136, 141–2; Turner, op. cit., 35, 41–2; J. A. Giles, *Hist. Par. and Tn Bampton*, Bampton, 1848, pp. 20, 75; Thirsk, 'Com. Flds', 26, 29; Hogan, art. cit., 125, 130; Barratt, op. cit., i, 49sqq., 96sqq., 115sqq., B. K. Roberts in Baker and Butlin, op. cit., 200; Postgate, ibid., 297; Shak. Bpl., Man. Docs, Ct Bk Alveston and Tiddington

21 Oct. 5 Jas, 19 Apr. 6 Jas; Ct R. id. 4 Oct. 1728; Oxon RO, Wi.x/34 2nd p.98; Glos. RO, D.184/M18 mm. 1, 12, 14, 19; M19 m. 12; M20 m. 14; Worcs. RO, B.A. 2358 no.131a; B.A. 351 bdl. 8 no. 1; PRO, Chanc. Proc. ser. i, Jas B12/68 m.3.

277 Ibid.; Marshall, *Glos.*, i, 68–9.

278 My *Ag. Rev.*, 92; Havinden in E. L. Jones, op. cit., 73–4.

279 Postgate in Baker and Butlin, op. cit., 298; Vancouver, *Cams.*, 125–7, 135sqq., 142–3, 145–7, 149, 151, 158sqq., 191; BL, Sloane MS 3815 fo. 141 (p. 319); Northants. RO, F(M) Coll. Misc. Vol. 99; cf. *Trans. Cambs. and Hunts. Archaeol. Soc.* 1904 i, 143, 150sqq.

280 A. Harris, *Rural Landscape E. R. Yks. 1700–1850*, 1961, pp. 41–2; *Open Flds* 9; Thirsk, 'Com. Flds', 27; Sheppard in Baker and Butlin, op. cit., 152, 170; PRO, D.L., M.B. 119 fos. 133v–4; Wards, Feodaries' Survs, bdl. 46 (Wilts.) unstd survs, Th. Flower/Beanacre; my *Ag. Rev.*, 143–4; cf. H. L. Gray, op. cit., 475–6, 505; J. E. Jackson, op. cit., 39, 42, 114–15, 137.

281 My *Ag. Rev.*, 92; Postgate in Baker and Butlin, op. cit., 297; M. Hollings, *Red Bk Worc.* Worcs. Hist. Soc. 4 pts, 1934–50, p. 171; Homans, op. cit., 57–8; Hallam in Finberg and Thirsk, op. cit., ii, 272; Raftis, ibid., 327–8; Fox, 'Alleged', 529–30, 532; cf. H. L. Gray, op. cit., 78, 453–5, 457, 469, 475, 480, 488.

282 Ibid., 74, 480, 486, 502; Ballard, 'Open Flds', 31; 'Tackley', 31; my *Ag. Rev.*, 65–6; Dyer in Finberg and Thirsk, op. cit., ii, 369; Turner, op. cit., 27; Plot, *Oxon.*, 243; A. Savine, *Engl. Mons on Eve Dissolution* Oxf. Studs in Soc. and Leg. Hist. i, 1909, p. 174; C. D. Ross and M. Devine, *Cartulary Cirencester Abb.*, 3 vols, London and Oxford, 1964–77, pp. 293, 345, 351–2, 774, 802–3, 805–6, 830–1, 883, 892, 897, 945, 971–2, 983–6; Hollings, op. cit., 279, 295, 317, 327, 354, 368; PRO, St. Ch. Proc. Jas 18/12 m. 9; 108/2 m. 2; Wards, Feodaries' Survs bdl. 46 (Wilts.) unstd Survs, Mich. Cooke, Broad Blunsdon; Shak. Bpl. Leigh Coll. Ct R. Adlestrop 25 Oct. 1725; Longborough 5 May 1691; Oxon. RO, Misc. M.I/1; Dashwood Coll. VIII/xxxiv, valn Harrison's fm; M. Whitfield, 'Med. Flds S. E. Soms.', *Soms. Archaeol. and Nat. Hist.* 1981 cxxv, 21; cf. J. E. Jackson, op. cit., 105.

283 J. Smyth, *Descron Hund. Berkeley* (Berkeley MSS iii) ed. J. Maclean, Gloucester, 1885, p. 4; BL, Add. MS 6027 fos. 32 (31)v–33 (32), 36 (35); PRO, Exch. T.R., Bk 157 p. 61 (fo. 32); D.L., M.B. 117 fo. 77; Glos. RO, D.444/M1, Ct R. Sevenhampton 1573.

284 Glos. RO, D.444/M2, Ct R. 2 Oct. 6 Chas.

285 BL, Add. MS 36875 fo. 29.

286 Ibid., fos. 29, 31v; Add. R. 18517; PRO, Exch. Deps by Commn 31 Eliz. Hil. 25 complt; St. Ch. Proc. Jas 108/2 mm. 1, 2; Oxon. RO, DIL.II.a.2; DIL.II.w.75, 13 Apr. 16 Chas.

287 Oxon. RO, Leigh Coll. Ct R. Adlestrop 25 Oct. 1725; Bodl. Hearne's Diaries 158 p. 28.

288 Bodl. Aubrey MSS 2 fos 83–4; 3 fo. 25v (p. 15); Worcs. RO, Cons. Ct Wigorn inv. 1566/73; J. Leland, *Itin.* ed. L. T. Smith, 5 vols, 1906–10, iii, 102; Marshall, *Glos.*, ii, 57.

289 Ibid., i, 328; ii, 36; Plot, *Oxon.*, 153, 242–3; Hone, op. cit., 175; Ballard, 'Tackley', 38; Ross and Devine, op. cit., 351; Morton, op. cit., 56;

Havinden in E. L. Jones, op. cit., 73–4; H. M. Colvin, *Hist. Deddington, Oxon.*, 1963, p. 89; Vinogradoff, *Villainage*, 226–7; Webb, op. cit., pt i, 79–81; BL, Add. MSS 23150 fo. 23; 36585 fos 164, 262, 284v, 300, 314; Birm. Ref. Lib. MS 168002, 16 Oct. 1679; 168163, 12 Oct. 1702; Oxon. RO, DIL.II.w.75, 13 Apr. 16 Chas; Dashwood Coll. VIII/xxxiii, Ct R. Duns Tew 5 Apr. 15 Jas; Shak. Bpl. Leigh Coll. Ct R. Adlestrop 10 Oct. 1701, 21 Apr. 1738; Longborough 24 Apr. 1712, 13 Apr. 1680, 30 Sept. 1678, 3 Oct. 1673; Glos. RO, D.745 Ct files Bisley, 1 Mar. 1675; Worcs. RO, Cons. Ct. Wigorn invs 1545/75; 1546/22; 1563/9, 81; cf. Hallam in Finberg and Thirsk, op. cit., ii, 345.

290 Straton, op. cit., 267; M. Chibnall, *Sel. Docs Engl. Lds Abb. Bec* R. Hist. Soc. Camd. 3rd ser. 1951 lxxiii, 90, 135; Titow, *Engl. Rural Soc.*, 146; Fussell, op. cit., ix; *V. C. H. Wilts.*, iv, 14, 15; PRO, Exch. L.R., M.B. 187 fo. 226; A.O. Parl. Survs Wilts. 43 fos 3–6; Wards, Feodaries' Survs bdl. 46 (Wilts.) unstd survs: Nic. Curtis/Ludgershall, Wm Noyes/Urchfont, Wm Upton/Burbage; Wilts. RO, Acc. 283 Liber Supervisus Amisburie; Savernake Coll. box 6 no. W. 19a Easton surv. c. 1760, p. 2; H. L. Gray, op. cit., 453, 502; my *Ag. Rev.*, 46.

291 Ibid., 61–2; cf. H. L. Gray, op. cit., 475–6, 505; Miller in Finberg and Thirsk, op. cit., ii, 399.

292 My *Ag. Rev.*, 52, 73, 83; PRO, Wards, Feodaries' Survs bdl. 46 (Wilts.) unstd survs, Jn Bryan(t)/S. Marston; cf. R. A. Pelham, 'Ag. Geog. Chich. Estes in 1388', *Suss. A. C.* 1937 lxxviii, 198.

293 My *Ag. Rev.*, 46, 52, 62–3, 76, 83, 142, 147; Brandon, art. cit., 62, 65; Otway-Ruthven, art. cit., 6–9; A. J. Roderick, 'Open-Fld Ag. in Herefs. in late M. A.', *Trans. Woolhope Naturalists' Fld Club* 1948–51 iii, 55sqq.; F. H. Manley, 'Customs Man. Purton', *W. A. M.* 1917–19 xl, 116–17; Bodington, art. cit., xli, 28; Miller in Finberg and Thirsk, op. cit., ii, 399; Jack, ibid., 414sqq.; Thirsk, 'Com. Flds', 26; Titow, *Engl. Rural Soc.*, 145–6; *V. C. H. Wilts.*, iv, 14, 15; H. L. Gray, op. cit., 462, 465–6, 468–9, 503, 505; PRO, Req. Eliz. 74/95; Devizes Mus., Wm Gaby His Booke, pp. 30, 42, 60–1, rev. pp. 15, 30; Berks. RO, D/EBt/E.28; Wilts. RO, Keevil and Bulkington Ct Bk 44 Eliz.–2 Chas, 16 Apr. 11 Jas, 2 May 15 Jas; Savernake Coll. box 6 no. W. 19a Easton surv. *c.* 1760, p. 2; Ct Bk Collingbourne Ducis 25 Apr. 1593; BL, Sloane MS 3815 fo. 141 (p. 319); Oxon. RO, Dashwood Coll. VIII/xxxiv, valn Harrison's fm.

294 Ibid.; Wilton Ho. Ct R. 1666–89, Stoke Verdon 19 Sept. 1677; Ct R. 1689–1723, box ii, vol. i, p. 82; box iii, vol. iii, pp. 61, 69; vol. iv, p. 68; Survs Mans 1631 vol. i, Fugglestone, fo. 6 (7); Ebbesborne Ct Bk beg. 1743, ao 1744; Foffont Ct Bk beg. 1743, ao 1743; Berks. RO, D/EBt/E.28; Brandon, art. cit., 65; my *Ag. Rev.*, 46, 52, 64.

295 Ibid., 147.

296 Ibid., 46; PRO, Exch. A.O., Parl. Survs, Wilts. 43 fo. 4.

297 M. R. Postgate, 'Hist. Geog. Breckland, 1600–1850', ts. thes. M.A. London Univ. 1960, pp. 99, 111, 117–18; art. cit., 83, 88–91, 93; in Baker and Butlin, op. cit., 300–2; H. L. Gray, op. cit., 311–12; E. M. Leonard, *Early Hist. Engl. Poor Relief*, Cambridge, 1900, p. 334; BL, Add. R. 16549, 63505, 63591; Sfk RO (Ips.) V.11.2.1.1; PRO, Exch. A.O., M.B. 419 fo. 59; D.L., R.&S. 9/5, 6; Req. Eliz. 65/52; Chanc. Proc. ser. i, Jas, L.15/7; W.11/60 compl.

298 My *Ag. Rev.*, 73.

299 Ibid., 145; Miller in Finberg and Thirsk, op. cit., ii, 400.

300 Lancs. RO, DD.F.1649.

301 Lancs. RO, DD.K.1542/2; and see J. Singleton, 'Infl. Geog. Facs on Dev. Com. Flds Lancs.', *Trans. Hist. Soc. Lancs. and Ches.* 1964 (1963), cxv, 35–6.

302 Owen, op. cit., pt i, 59, 62; pace my *Ag. Rev.*, 157.

303 As nn. 142, 201.

304 C. V. Goddard, 'Customs Man. Winterbourne Stoke 1574', *W. A. M.* 1905–6 xxxiv, 212; Wilts. RO, Acc. 283 Liber Supervisus Amisburie; Bodl., Aubrey MSS 2 fo. 84; cf. Straton, op. cit., 311.

305 Marshall, *Glos.*, i, 65–6, 71–3; Roberts in Baker and Butlin, op. cit., 201; J. A. Yelling, *Com. Fld and Enclo. in Engld 1450–1850*, 1977, p. 161.

306 My *Ag. Rev.*, 174; Brandon in Finberg and Thirsk, op. cit., ii, 320.

307 PRO, Exch. A.O., Parl. Survs Northants. 32 fos 3, 4, 6, 8, 12–14, 18, 22, 26, 29, 37, 39, 42sqq., 49, 61, 68, 73, 78; Northants. RO, F(M) Coll. Misc. Vols 48 fos 47–8, 61vsqq.; 99; 437 p. 18; Yelling, op. cit., 161; Morton, op. cit., 14.

308 Ibid.

309 Marshall, *E. Dept*, 6; J. Sinclair, *Ag. Hints*, n.d., pp. 14, 15; T. Stone, *Rev. Corr. Ag. Surv. Lincs. by Art. Young Esq.*, 1800, pp. 96–7; Vancouver, *Cambs.*, 154, 159–60, 217; Norden, *Surveiors Dialogue* 213; C. V. Collier, 'Stovin's MS', *Trans. E. R. Archaeol. Soc.* 1905 (1904) xii, 50; C. Jackson, 'N. fr. Ct R. Man. Epworth in Co. Lincs.', *Reliquary* 1882–3 xxiii, 91; BL, Add. MS 37521 fos 30, 38 recte (58 erron.), 43v; Northants. RO, F(M) Coll. Misc. Vol. 99.

310 Marshall, *Glos.*, i, 65–6, 72–5, 98–9, 122; Turner, op. cit., 35; Gooch, op. cit., 96; W. James and J. Malcolm, *Gen. View Ag. Co. Sy*, 1794, pp. 45, 48, 59; *Bucks.*, 27; Morton, op. cit., 14; R. C. Richardson in Finberg and Thirsk, op. cit., v, pt i, 261; Straton, op. cit., 311; my *Ag. Rev.*, 174.

311 Northants. RO, F(M) Coll. Misc. Vol. 437 pp. 3, 7, 16, 18; Maps 1000, 1004; Worcs. RO, B.A. 351 bdl. 8 no. 1; Finberg and Thirsk, op. cit., v, pt i, 257; Spufford, *Contrasting Communs*, 96; Hollings, op. cit., 1–3, 39, 171; Hogan, art. cit., 125, 130.

312 PRO, L.R., M.B. 196 fos 130sqq.; 208 fos 3sqq.; James and Malcolm, *Bucks.*, 21.

313 Beds. RO, S. 20.

314 PRO, Exch. L.R., M.B. 222 fos 201sqq.; 228 fos 3sqq.; BL, Add. R. 54166; Shak. Bpl. Leigh Coll. Ct Bk Stoneleigh p. 43; Ct files id. 6 Oct. 1692; Birm. Ref. Lib. MS 378173.

315 Leics. RO, Arch. inv. 1669/32.

316 Ches. RO, Vernon (Warren) Coll. Ct R. nos 1, 8, 9.

317 PRO, Exch. A.O., Parl. Surv. Northants. 32 fos 3 (1)sqq.; Staffs. RO, Hand Morgan Coll., Chetwynd MSS, file O, Ct R. Ingestre and Salt 21 Apr. 27 Eliz.

318 Barratt, op. cit., i, 26sqq.

319 Vancouver, *Cambs.*, 147; Glos. RO, D. 184/M24, p. 139; Worcs. RO, B.A. 351 bdl. 8 no. 1, p. 19; Shak. Bpl., Man. Docs, Ct R. Alveston and Tiddington 4 Oct. 1728.

320 T. Davis, op. cit., (1794) 9.

321 Wilts. RO, Savernake Coll. box 6 no. W.19a Easton surv. *c.* 1760 p. 2; Maskelyne and Manley, art. cit., 475.

322 PRO, Exch. L.R., M.B. 187 fo. 226.

323 Berks. RO, D/EHy M26.

324 PRO, Exch. A.O., Parl. Surv. Wilts. 43 fos 3sqq.; J. E. Jackson, op. cit., 29.

325 Wilts. RO, Acc. 283 Liber Supervisus Amisburie.

326 C. Dyer, op. cit., 68, 322; W. H. Hale, *Registrum sive Liber Irrotularius et Consuetudinarius Prioratus Beatae Mariae Wigorniensis* Camd. Soc. 1865 xci, fos 82, 86b, 87a; Hollings, op. cit. 1–3, 39, 151, 243, 263; PRO, Chanc. Proc. ser. i, Jas B. 12/68 m. 3.

327 Ibid. S. 29/7 mm. 1, 2; Birm. Ref. Lib. MS 344742; Lincs. RO, Monson Dep., Newton Pprs 7/12/82.

328 Fox, 'Alleged', 538–40, 543–4.

329 Ibid., 529–30, 532–4, 538sqq.; Titow, *Engl. Rural Soc.*, 39; H. L. Gray, op. cit., 74sqq., 80–1.

330 Ibid., 73, 126–7, 493–4; Ballard, 'Tackley', 31–2; 'Open Flds', 23, 26–7, 31; Oxon. Archaeol. Soc. *Rep.* 1911 (1912) 138; Colvin, op. cit., 87–8; Barratt, op. cit., i, 59sqq., 78sqq.; M. A. Havinden, 'Rural Econ. Oxon. 1580–1730', ts. thes. B. Litt. Oxford Univ. 1961, p. 268; 'Ag. Prog.', 72–3; R. W. Jeffery, *Mans and Advowson Gt Rollright* Oxon. Rec. Soc. 1927 ix, 23; BL, Add. MSS 36585 fo. 164; 36649 fos 4, 5; Add. R. 9288 mm. 1–4; 9289 ao 1659; Shak. Bpl. Leigh Coll. Ct R. Longborough 28 Mar. 1625, 5 May 1691; Ct R. Adlestrop 25 Oct. 1725.

331 Ibid.; Havinden, 'Rural Econ.', 268.

332 Northants. RO, Ct R. Aynho 8 Oct. 1639, 1 Oct. 1652, 9 Apr. 1680; Fox, 'Alleged', 540; J. Ritchie, *Reps Cs decided by Francis Bacon in Ct Chanc. (1617–21)*, 1932, pp. 183–4; Oxon. RO, Dashwood Coll. VIII/xxxiv, valn Harrison's fm.

333 Ibid.; Misc. M.I/1, 10 Apr. 24 Chas.

334 Ballard, 'Open Flds', 26; Shak. Bpl. Leigh Coll. Ct R. Adlestrop 10 Oct. 1701.

335 Oxon. RO, DIL.IV.a.78, iii, v; 89, 29 Apr. 1649.

336 Webb, op. cit., pt i, 81; Northants. RO, Ct R. Aynho 8 Oct. 1639, 1 Oct. 1652, 5 Apr. 1672, 9 Apr. 1680; Shak. Bpl. Leigh Coll. Ct R. Adlestrop 25 Oct. 1725, 21 Apr. 1738, 9 Feb. 1741–2.

337 As nn. 262–71, 273, 281, 287, 309.

338 J. Childrey, *Britannia Baconica*, 1660, p. 103; Folkingham, op. cit., 42; Spufford, *Contrasting Communs*, 95; 'Rural Cambs.', 106; BL, Harl. MS 304 fo. 75v; my *Ag. Rev.*, 59; R. Holinshed and W. Harrison, *Chrons*, vols i and ii, 1586, p. 233.

339 Ibid., 232–3; C. Howard in *Phil. Trans. R. S. Lond.* ed. C. Hutton, G. Shaw and R. Pearson, 18 vols, 1809, ii, 423–4; *V. C. H. Herts.*, iv, 214; *Ess.*, ii, 359–60, 362, 364; T. Fuller, *Hist. Worthies Engld* (1662) ed. P. A. Nuttall, 3 vols, 1840 i, 222, 492; R. Blome, *Brit.* 1673, p. 95; J. Laurence, op. cit., 115; Folkingham, op. cit., 30–1, 35; B[lagrave], op. cit., 217–19; J. Norden, *Speculi Britanniae Pars: Historical and Chorographical Descron Co. Ess.*

(1594) ed. H. Ellis, Camd. Soc. 1840 [ix], 8; J. Gerarde, *Herball*, 1597, pp. 123–4; Hull, op. cit., 20; Spufford, 'Rural Cambs.', 106; Blith, op. cit., 244–5; BL, Harl. MS 304 fo. 75v; Add. R. 27170; PRO, Exch. K. R. Deps by Commn 4 Jas Mich. 25; Northants. RO, Cons. Ct Petriburg, Cams. Bds and Invs 1662: Jas Bankes sr/Stapleford, Alice Barber wid./Fulbourn, Jn Haylocke/Sawston.

340 Bodl. Hearne's Diaries 158 p. 28; BL, Add. MS 37682 fos 8, 9v, 10.

341 Orwin, op. cit., 161, 163, 165, 173, 176, 180; (1954) 55, 61, 143.

342 *V. C. H. Leics.*, ii, 212–13, 220; R. H. Hilton, *Econ. Dev. some Leics. Estes in 14th and 15th cents*, Oxford, 1947, pp. 52–3.

343 As nn. 420–2, 431.

344 Wilts. RO, Savernake Coll. Ct Bk Collingbourne Ducis 25 Apr. 1593.

345 Wilts. RO, Acc. 7, Enford Ct Bk fo. 107 (47).

346 PRO, Sp. Colls, Ct R., G.S. bdl. 209 no. 31, 10 Oct. 1598.

347 My *Ag. Rev.*, 93–6; as nn. 420–2, 431.

348 Yelling, op. cit., 165–8, 170, 201; Harris, *Open Flds*, 7, 10; Havinden, 'Ag. Prog.', 70; Atwell, op. cit., 101–2; Worcs. RO, B.A. 54/E80, Ct R. Kempsey 31 Mar. 1656; my *Ag. Rev.*, 278, 286, 289sqq.

349 Ibid., 289sqq.

350 As nn. 118–23.

351 Marshall, *Midl. Dept*, 401; *E. Dept*, 49; James and Malcolm, *Bucks.*, 30; W. Mavor, op. cit., 137; T. Davis, op. cit. (1794) 84; (1813) 46; Ruston and Witney, op. cit., 177; Beds. RO, L.4/333.

352 T. S. Ashton, *Ind. Rev.*, 1948, p. 60; Gonner, op. cit., 57; W. Notestein, F. H. Relf and H. Simpson, *Commons Debates 1621*, 7 vols, New Haven, 1935, ii, 173; Stat. 18 Geo. 3 c. 81.

353 My *Ag. Rev.*, 43–4, 46, 52, 62–3, 81, 100, 143, 174, 240.

354 Ibid., 43–4.

355 T. Davis, op. cit. (1794) 15, 17; Worlidge, op. cit., 66; Naish, op. cit., 61; Ault, *Husb.*, 26.

356 Wilts. RO, Acc. 283, Ct Bk Amesbury Erledom 15 Mar. 28 Eliz.

357 Wilton Ho. Ct Bks Var. Pars 1633–4 fo. 46v.

358 Marshall, *S. Cos*, ii, 319, 350–1.

359 Wilts. RO, Hyde Fam. Docs, Cts Sir Laurence Hyde, Heale 23 Sept. 1629.

360 Gras, *Engl. Vil.*, 510; Baigent and Millard. op. cit., 323, 325–6, 333, 350; Berks. RO, D/EBt.E28; D/EC.M151; D/EHy/M32; Wilton Ho. Ct R. 1689–1723 box ii, vol. i, p. 90; box iii, vol. ii, pp. 43, 47.

361 Cn Bennett, 'Ords Shrewton', *W. A. M.* 1887 xxiii, 36–9; Wilton Ho. Ct Bk Var. Pars 1633–4 fo. 6; Wilts. RO, Hyde Fam. Docs, Ct Sir Laurence Hyde 7 Apr. 1609; Acc. 283, Ct Bk Amesbury Erledom 15 Mar. 28 Eliz.; BL, Harl. MS 6006 fo. 52.

362 T. Davis, op. cit. (1794) 15.

363 Wilts. RO, Acc. 283, Ct Bk Amesbury Erledom 15 Mar. 28 Eliz.

364 C. V. Goddard, art. cit., 212.

365 BL, Add. MS 23152 fos 14v, 15.

366 Wilts. RO, Savernake Coll. Ct R. All Cannings 27 Apr. 16 Jas.

367 BL, Harl. MS 6006 fos 17v, 52.

368 Wilts. RO, Hyde Fam. Docs, Ct Bn Brigmilston and Milston, Sept. 1610.

369 Straton, op. cit., 311; *V. C. H. Wilts.*, iv, 14, 27.

370 Wilton Ho. Ct Bk Var. Pars 1633–4 fo. 47.

371 Gras, *Engl. Vil.*, 592–3.

372 Marshall, *S. Cos*, ii, 367; G. M. Cooper, art. cit., 232; J. C. K. Cornwall, 'Agrarian Hist.', 101; 'Fmg in Sx', *Suss. A. C.* 1954 xcii, 52–3; Thomas-Stanford, op. cit., 7, 8; Godfrey, op. cit., 61–2, 225; Northants. RO, Wmld Coll. 5, ix, surv. Rodmell.

373 BL, Add. R. 28264, 16 Apr. 3 Chas, 4 Apr. 9 Chas.

374 G. M. Cooper, art. cit. 240–1; Marshall, *S. Dept*, 500, 504.

375 Wilts. RO, Keevil and Bulkington Ct Bk 44 Eliz. – 2 Chas, 1 Apr. 10 Jas, 16 Apr. 11 Jas, 28 Apr. 18 Jas; BL, Add. MS 37270 fos 114, 136v; Bowood Ho. Surv. Bremhill 1629.

376 Seebohm, op. cit., 12, 450; Austin, op. cit., i, 259–60, 271, 273; ii, 272–3; Whiteman, op. cit., 99, 101, 103; BL, Add. MS 27977 fos 3, 358; Bucks. RO, Ct R. Pitstone P.24/2, 7 Oct. 7 Chas; P.24/3, 14 Apr. 1651, 12 Oct. 1677; I.M.6/12 Ct Bk Ivinghoe p. 2; Herts. RO, Ashridge Coll. 723; Cashiobury and Gape Coll. 9464 29 Mar. 17 Jas; 9479, 9 Apr. 1634; Pym Coll. 19271 Ct. R. Norton 28 Apr. 1606, 11 Apr. 1608; Misc. MSS 2665, 25 Sept. 22 Jas; 7052, 1 Dec. 5 Chas; 41674 fo. 13; 65810 pp. 143sqq.

377 My *Ag. Rev.*, 64, 67; Glos. RO, D.158/M1 Ct R. Bourton-on-the-Water 7 Apr. 16 Eliz.; Hilton, *Med. Soc.*, 108.

378 Ibid.; my *Ag. Rev.*, 96; Marshall, *Midl. Dept*, 181, 201, 268, 327, 383–4, 401, 423–4, 428, 532–3, 560, 563–4, 594, 603; *Midl. Cos*, ii, 225; Lisle, op. cit., 120; Morton, op. cit., 480, 482; Plot, *Oxon.*, 239; *Staffs.*, 340; W. Pitt, *Gen. View Ag. Co. Northampton*, 1809, pp. 82, 176; *Gen. View Ag. Co. Leicester*, 1809, p. 72; Crutchley, op. cit., 15; Batchelor, op. cit., 358, 389, 445–6, 537, 560–1; Parkinson, *Ruts.*, 49; J. Donaldson, *Gen. View Ag. Co. Northampton*, Edinburgh, 1794, p. 29; J. H. Blundell, *Toddington: its Ann. and People*, Toddington, 1925, p. 65; S. P. Potter, *Hist. Wymeswold*, 1915, p. 89; BL, Add. R. 26836 Glatton cum Holme 1 Apr. 19 Eliz., 4 Apr. 20 Eliz.; Spufford, 'Rural Cambs.', 12; Emmison, *Jacobean Hsehld Invs*, Beds. Hist. Rec. Soc. 1938 xx, 70–1; Fitzherbert, *Husbandrie*, fo. 11v (recte) 9v (erron.); *Surveyinge*, fos 42v–3; T. Pape, *Newcastle-under-Lyme in Tu. and early Stuart Times*, Manchester, 1938, p. 100; W. Ellis, *Compleat Syst. Experienced Improvements made on Sheep, Grass-lambs and House-lambs*, 1749, pp. 219–21, 226.

379 Ibid., 131–2; Marshall, *Midl. Dept*, 201, 383–4, 533, 594–5; Pitt, *Leics.*, 72; BL, Add. MS 36908 fos 45, 52v, 55v; PRO, Chanc. Proc. ser. i, Jas T.13/46; Sp. Colls, Ct R., G.S. bdl. 194 no. 55, 16 Oct. 4 Eliz.; 195 no. 78 m. [4d.]; Shak. Bpl. W. de B. Coll. 1194, 6 Apr. 11 Chas; Throckmorton Coll. Ct R. Weston Underwood 23 Oct. 1579; Overley 30 Sept. 1616; Leigh Coll. Ct R. Stoneleigh 2 Oct. 20 Eliz., 27 Mar. 42 Eliz.; pains and ords 1597; Deene Ho. Brudenell MSS A.xviii.15; O.xxiv.1, Ct Bk Thistleton 10 Apr. 15 Eliz.; Northants. RO, I(L) Coll. 128, 25 Oct. 11 Eliz., 13 Apr. 13 Eliz.; 3545 fo. 9; Ecton Coll. 1191; Wmld. Coll. box 5 pcl v no. 1, Ct R. Tansor 13 Sept. 19 Jas; box 4 pcl xvi no. 5, surv. Nassington; Mont. Coll. Northants. box 16 no. 98 Ct Bk L. Oakley 14 Jan. 17 Eliz., 14 Sept. 1648; box 7 no. 66/5 Ct files

Weekley 8 Oct. 35 Eliz.; 66/6 Ct R. id. 16 Oct. 20 Jas; box 17 no. 159 Ct R. Broughton 30 Oct. 24 Eliz.; no. 160 id. 27 Apr. 17 Eliz.; 30 Oct. 24 Eliz.; 18 no. 160 Ct R. Barnwell 9 Dec. 5 Chas; box 14 no. 12 Ct R. Brigstock 2 May 21 Eliz.; Misc. Led. 145 pp. 157, 214, 371, 460, 535; 146 pp. 12, 174; Rushden Ct R. presentmts 17–20 Eliz.; Cambs. RO, L.1/112 Ct R. Long Stanton 3 Oct. 1654; Beds. RO, TW 10/2/9 Clapham byelaws 14 Apr. 8 Chas, 4 Apr. 1659; Ault, *Fmg*, 140; *Husb.*, 26, 87.

380 Ibid., 26, 88; Cambs. RO, L.1/112 Ct R. Long Stanton 3 Oct. 1654; L.19/20 Ct R. Haslingield 3 Apr. 24 Chas; War. RO, M.R. 14; BL Add. MS 36908 fos 14v, 45, 55v, 63v; Add. R. 44579, 44649, 49700; PRO, Exch. K.R. Deps by Commn 14 Jas Mich. 37; 13 Chas Mich. 48 ex parte quer. art. 6; Chanc. Proc. ser. i, Jas T.13/46; DL, Ct R. bdl. 82 no. 1133 m. 28; 106 no. 1529 mm. 1, 11; 1531 m. 5; 1532 mm. 1, 3; 1533 m. 2; 1535 mm. 5, 6d.; Sp. Colls, Ct R., G.S. bdl 195 no. 36 mm. 1d., 2; 79 mm. 1d., 7d.; 84 m. 1d.; Shak. Bpl. Leigh Coll. Ct R. Stoneleigh 6 Apr. 1562, 30 Sept. 1574, 30 Sept. 1596, 1 Apr. 17 Jas, 5 Oct. 1693; Ct files Sowe als Walsgrave 23 Apr. 1663; W. de B. Coll. 703b, 22 Oct. 1606; 1194; Hood Coll. Ct R. Stivichall 23 Oct. 1710, 3 Oct. 1724; Throckmorton Coll. Ct R. Weston Underwood 23 Oct. 1579; Upton in Haselor 20 Oct. 1692; Oversley 22 Oct. 1646, 13 Oct. 1762; Northants. RO, Mont. Coll. War. old box 6 no. 14 Ct R. Newbold 27 Oct. 1707; Northants. box 18 no. 160 Ct R. Barnwell 9 Dec. 5 Chas; box 7 no. 66/6 Ct R. Weekley 16 Oct. 20 Jas; box 17 no. 160 Ct R. Broughton 23 Mar. 4 Ed. 6; Misc. Leds 129 Polebroke 27 Oct. 1727; 132 Long Lawford 9 Oct. 1721; 145 pp. 41–2, 99, 371; Grafton Coll. 3438a; Ecton Coll. 1191; Higham Ferrers Ct Pprs, Burgess Ct R. Fld Ords 1696–1725: 28 Oct. 1711, 23 Oct. 1713, 28 Apr. and 24 Oct. 1719, 2 Apr. 1725; Raunds Ct R. 13 Eliz.; Deene Ho. Brudenell MSS I.v.63; O.xxiv.1, Ct Bk Thistleton 10 Apr. 15 Eliz., 15 Apr. 36 Eliz.; Deene Oct. 18 Eliz.; Cranoe 28 Apr. 16 Eliz.; Leics. RO, DE.40/40 Ct Bn Ullesthorpe 20 Aug. 1684, 8 Nov. 1687, 1 Oct. 1690; Beaumanor Coll. Ct Bk 'Mr Robt Pilkington Steuarte' fo. 9; Ct R. and files Barrow etc.: Barrow 22 Oct. 1717; Quorndon 26 Oct. 1756; Ches. RO, Vernon (Warren) Coll. Ct R. 3; Pape, op. cit., 100; Morton, op. cit., 480, 482; Plot, *Oxon.*, 239; Ellis, *Compl. Syst.*, 220–1, 224; Lisle, op. cit., 120, 324–5; Fitzherbert, *Surveyinge*, fo. 43; *Husbandrie*, fos. 8, 11v (recte) 9v (erron.), 13v, 18; F. P. Verney, *Mem. Verney Fam. during Civ. War*, 3 vols, 1892–4, iii, 135; Marshall, *Midl. Cos*, ii, 226; *Midl. Dept* 383, 401, 423, 594, 598; Pitt, *Northants.*, 82; Batchelor, op. cit., 389.

381 My *Ag. Rev.*, 114; Marshall, *Glos.*, i, 209; *W. Dept*, 417–18; Turner, op. cit., 42; Glos. RO, D.184/M9 fos 1vsqq.; M18 mm. 1, 14, 19; M24 pp. 58sqq., 69sqq.; D.326/E1 fo. 48v; Cons. Ct Gloc. invs: 1615/1; 1633/9; PRO, Exch. L.R., M.B. 228 fo. 296 (8); Shak. Bpl. Man. Docs, Ct R. Alveston and Tiddington 11 Apr. 18 Chas 2; Ct Bk. id. 15 Apr. 9 Jas; Stratford Coll. Misc. Docs vol. i nos 11, 17, 19, 30–1, 65, 77; vol. v no. 32; vol. vii no. 144; Wheeler Pprs vol. i fo. 67; Worcs. RO, Cons. Ct Wigorn invs: 1545/285, 300; 1546/37, 103, 105; 1563/18, 42; 1566/118; 1592/28; 1596/61, 74; 1612/1, 6; 1615/236; 1617/134; 1633/105; 1638/151.

382 My *Ag. Rev.*, 121, 143–4.

383 Ibid., 118–19, 121, 143; Straton, op. cit., 505sqq.; BL, Harl. MS 71 fos 45 (56), 49 (60); PRO, Exch. A.O., Parl. Survs: Dors. 9 fos 6, 9; Soms.

38 fo. 7; L.R., M.B. 207 fos 20–1 (18, 19); Soms. RO, DD/MI Acc. C. 186 box 6 Ct R. Q. Camel 12 Oct. 1696, 24 Oct. 1698, 3 Oct. 1700.

384 Lisle, op. cit., 325; my *Ag. Rev.*, 123.

385 Ibid., 147; J. Beale, *Herefs. Orchards: patt. f. all Engld*, 1657, pp. 51, 55; Stat. 4 Jas c. 11.

386 My *Agrarian Probs*, 128–9; *Ag. Rev.*, 43, 52–3; Godfrey, op. cit., 63, 65, 67, 69; Northants. RO, Wmld Coll. 5.ix, survs Northease and Iford, Rottingdean, Rodmell.

387 W.O. Ault, *Ct R. Abb. Ramsey and Hon. Clare*, New Haven, 1928, pp. 209, 274; N. Neilson, *Econ. Conds on Mans Ramsey Abb.*, Philadelphia, 1899, p. 37; J.E. Jackson, op. cit., 140; M.D. Lobel, *Boro. Bury St Edms*, Oxford, 1935, pp. 20, 26; Page, *Crowland Abb.*, 50, 107, 446; *V. C. H. Wilts.*, iv, 26–7; G.M. Cooper, art. cit., 234; Campbell, 'Regional Uniqueness', 23–5; M. Bailey, 'Sand into Gold: evol. foldcourse syst. in W. Sfk, 1200–1600', *Ag. Hist. Rev.* 1990 xxxviii, 53; C.G.O. Bridgman, 'Burton Abb. 12th cent. Survs', *Colls Hist. Staffs.* 3rd ser. 1916 (1918), 212, 222, 230, 243–4; Rec. Commn, *Rotuli Hundredorum*, ii, 458; W.H. Hale, *Dom. St Pauls in Yr 1222*, Camd. Soc. 1858 lxix, 105; Douglas, op. cit., 25, 78; Vancouver, *Cambs.*, 123; M. Chibnall, op. cit., 70, 132, 172, 176–7, 182; W. Cunningham, 'Com. Rts at Cottenham and Stretham', *Camd. Misc. xii*, Camd. 3rd ser. 1910, xviii, 253; Spufford, *Contrasting Communs*, 129; Postgate in Baker and Butlin, op. cit., 318–19; as n. 369.

388 D. Woodward, op. cit., xxxi, 99, 104, 220–2; Lincs. RO, Heneage of Hainton Coll. 3/3; my *Ag. Rev.*, 62.

389 Ibid., 75, 77, 88; C. Parkin in M. Blomefield, *Essay twds Topog. Hist. Co. Nfk*, 5 vols, Lynn, 1775, v, 799; H.W. Chandler, *5 Ct R. Gt Cressingham in Co. Nfk*, 1885, p. 12; Douglas, op. cit., 25–6, 78–9; Postgate in Baker and Butlin, op. cit., 314sqq.

390 Ibid., 317; Marshall, *Midl. Dept*, 638; *E. Dept*, 424; [Hervey], op. cit., 6; J. Saltmarsh and H.C. Darby, 'Infld–Outfld Syst. on Nfk Man.', *Econ. Hist.* 1934–7 iii, 38–9; Spratt, op. cit., 229, 245; Ess. RO, D/DP, E25 fos 23, 53; PRO, Chanc. Proc. ser. i, Jas B.14/17; B.28/46 mm. 136–7; C.1/50 complt; C.6/24 complt; C.6/35 complt; C.16/29 mm. 1, 4; D.8/11 complt; L.15/7 complt and ans.; U.2/35 complt; W.11/60 complt; W.26/15 complt; Req. Eliz. 65/52; 102/45 complt; Exch. L.R., M.B. 201 fo. 93v (quot.); 255 fos 48, 54v; Nfk RO, D.&C. MSS, Parl. Surv. fos 111, 117.

391 Ibid., fo. 103; Spratt, op. cit., 249; Postgate in Baker and Butlin, op. cit., 314–15, 318; H.L. Gray, op. cit., 325sqq.; K.J. Allison, 'Wl Sup. and Worsted Cl. Ind. in Nfk in 16th and 17th cents', ts. thes. Ph.D. Leeds Univ. 1955, pp. 32, 35–7; 'Sheep-Corn Husb. Nfk in 16th and 17th cents', *Ag. Hist. Rev.* 1967 v, 15, 16; J. Caius, *Ann. Gonville and Cai. Coll.* ed. J. Venn, C.A.S. 8vo Pubs 1904 xl, 310sqq.; PRO, Chanc. Proc. ser. i, Jas C.6/35 complt; C.M. Hoare, op. cit., 331.

392 Ibid., 332; Spratt, op. cit., 256–7; Allison, 'Sheep-Corn', 21; 'Wl Sup.', 44, 91, 98sqq., 161, 286, 296–7; PRO, Chanc. Proc. ser. i, Jas D.8/11 m. 1; Saltmarsh and Darby, art. cit., 38–9; Bailey, 'Sand into Gold', 45–8; *Marg. Econ.*, 66, 68, 76, 78; cf. BL, Add. MS 36745 fo. 11.

393 Bailey, *Marg. Econ.*, 68–71; Saltmarsh and Darby, art. cit., 37–9,

42; R.J. Hammond, 'Soc. and Econ. Circs Ket's Rebellion', ts. thes. M.A. London Univ. 1933, pp. 59, 61, 64; Allison, 'Wl Sup.', 177; 'Flock Man. in 16th and 17th cents', *Econ. Hist. Rev.* 2nd ser. 1958 xi, 99; PRO, Chanc. Proc. ser. i, Jas L.15/7 complt; W.11/60 complt; Exch. L.R., M.B. 201 fo. 93.

394 Postgate, 'Hist. Geog.', 157–8; C.M. Hoare, op. cit., 263; Slater, *Engl. Peasantry*, 332; H.L. Gray, op. cit., 326, 342–3; Gonner, op. cit., 17; Spratt, op. cit., 247–8, 255sqq.; cf. BL, Add. MS 36745 fo. 11.

395 Spratt, op. cit., 248; Hammond, op. cit., 57, 64–5; BL, Stowe MS 775 fos 42–3; Allison, 'Wl Sup.', 53, 248, 264–5.

396 Ibid., 181–2.

397 Nfk RO, Cons. Ct Norvic. inv. 1633/108.

398 BL, Add. R. 16551.

399 Allison, 'Wl Sup.', 206sqq.; 'Sheep-Corn', 100; Hammond, op. cit., 57, 64; F.W. Brooks, 'Suppl. Stiffkey Pprs', *Camd. Misc. xvi*, Camd. 3rd ser. 1936 lii, 40–2; Nfk RO, Nor. Lib. MS 1505 fos 1sqq.

400 BL, Stowe MS 775 fos 32vsqq.

401 BL, Add. R. 16548; PRO, Exch. L.R., M.B. 255 fo. 48; Ess. RO, D/DP E25 fo. 23; A. Simpson, 'E.A. Foldcourse: some qus', *Ag. Hist. Rev.* 1958 vi, 90; Hammond, op. cit., 59.

402 Ibid., 62–4; Bailey, 'Sand into Gold', 45–8, 50–1; *Marginal Econ.*, 74–5; cf. A. Simpson, *Wealth Gentry 1540–1660*, Cambridge, 1961, p. 189.

403 Hammond, op. cit., 80–2; PRO, Req. Eliz. bld. 65 file 52 compl.; M.M. Knappen, *Tu. Puritanism: chap. in hist. idealism*, Chicago and London, 1970, p. 408 (quot.); Allison, 'Sheep-Corn', 22–5.

404 Ibid.; 'Wool Sup.', 154sqq., 162sqq., 266, 275–6; PRO, Req. Eliz. 65/52 complt; 102/45 complt; Chanc. Proc. ser. i, Jas C.16/29 mm. 1, 4; St. Ch. Proc. Eliz. A.22/20.

405 Tusser, op. cit., 99, 136; cf. Postgate in Baker and Butlin, op. cit., 319–20.

406 A. Bland, P. Brown and R.H. Tawney, *Engl. Econ. Hist.: sel. docs*, 1925, pp. 247–8.

407 Ess. RO, D/DL M81 Ct Bk 1 fos. 42v, 57, 63, 81, 83, 90, 97–8, 109v–10, 111v, 118, 133v; T. Barrett-Lennard, '200 Yrs Este Man. at Horsford during 17th and 18th cents', *Nfk Archaeol.* 1921 xx, 61, 64sqq.

408 Spratt, op. cit., 238, 252, 254; Allison, 'Wool Sup.', 54–5, 57, 63–4; Postgate, 'Hist. Geog.', 161; in Baker and Butlin, op. cit., 320–1.

409 Holinshed and Harrison, op. cit., 233; Childrey, op. cit., 103; Folkingham, op. cit., 42; BL, Harl. MS 304 fo. 75v; as nn. 338–9.

410 PRO, Req. Eliz. 65/52.

411 PRO, Chanc. Proc. ser. i, Jas W.11/60 complt.

412 Ibid., L.15/7 complt.

413 Ibid., U.2/35 complt.

414 Ibid., B.14/17; my *Ag. Rev.*, 268.

415 Spratt, op. cit., 254.

416 PRO, Chanc. Proc. ser. i, Jas C.1/50.

417 Tawney, op. cit., 236.

418 Spratt, op. cit., 248.

419 My *Ag. Rev.*, 88, 301–2, 306; Davison, art. cit., 354; Postgate in Baker and Butlin, op. cit., 321–2.

420 Davies in id., 510sqq.; M. D. Harris, *Cov. Leet Bk 1420–1555* E.E.T.S. cxxxiv–v, cxxxviii, cxlvi 1907–13, pp. 633, 692, 719–20, 729sqq.; Cov. RO, City Recs, A.3 Leet Bk 30 Eliz.–1834, p. 15; Webb, op. cit., i, 166; Stark, op. cit., 74, 91, 93, 208; Lobel, op. cit., 20, 26; PRO, D.L., M.B. 122; S.P.D. Chas 475/70; R. C. Richardson and T. B. James, *Urb. Exp.: Engl., Scot. and W. tns, 1450–1700*, Manchester, 1983, pp. 54–5, 68, 80–1; D. M. Woodward, *Trade Eliz. Chester*, Hull, 1970, p. 107; Waylen, op. cit., 110, 120, 123; *V. C. H. Leics.*, iv, 102; my *Agrarian Probs*, 98–9.

421 Ibid., 24; Thornton, op. cit., 111–13; Wake, art. cit., 410–11; S. Elliott, art. cit., 157; W. O. Ault, 'Some Early Vil. By-laws', *Engl. Hist. Rev.* 1930 xlv, 208sqq.; 'Vil. By-laws by Com. Consent', *Speculum* 1954 xxix, 378sqq.; *Husb.*, 12, 26, 40–3, 51–4; *Ct. R. Ramsey*, 156, 200, 278–9; Stenton, *Docs Danelaw*, p. lxii; Neilson, *Terrier Fleet*, p. lxxviii; Raistrick, op. cit., 46; H. M. C. *Rep. on MSS Ld Middleton*, 1911, pp. 106–9.

422 Ibid.; G. H. Tupling, *Econ. Hist. Rossendale*, Chetham S. n.s. 1927 lxxxvi, 57; Raistrick, op. cit., 45–6; Cunningham, op. cit., 211; Potter, op. cit., 92; [W. H. Longstaffe & J. Booth], *Halmota Prioratus Dunelm.* Surtees Soc. 1889 lxxxii, pp. xii, xiii; W.Brown, *Yks. St. Ch. Proc.* Yks. Archaeol. S. Rec. Ser. 1909 xli, 81; Page, *Crowland Abb.*, 25–8, 164–5; W. G. Hoskins, 'Flds Wigston Magna', *T. L. A. S.* 1936–7 xix, 174–5; Vinogradoff, *Villainage*, 354–5; Hilton, *Med. Soc.*, 151, 155; my *Agrarian Probs*, 19sqq.; Deene Ho. Brudenell MS O.xxiv.1 Ct Bk Stonton Wyville 4 Dec. 19 Eliz.; Northants. RO, Old Par. Recs, Fld offrs' accts 1739; Daventry Coll. D.549b, Hellidon ords 27 Oct. 1740; Mont. Coll. Bucks. box P, pt 2, X.890 Ct files Dondon in Quainton 16 Apr. 36 Chas 2; Northants. old box 18 no. 160 Barnwell ords 14 Oct. 28 Eliz.; Ecton Coll. 1191 Ecton ords 14 Oct. 1743, nos 24–5; Misc. Leds 129, L. Oakley 13 Oct. 1721; 145, p. 98; Leics. RO, Beaumanor Coll. Ct R. and files Barrow etc. Quorndon 26 Oct. 1756; Shak. Bpl. Trevelyan Coll. Ct R. Snitterfield and Bearley 27 Oct. 1713; Beds. RO, TW 10/2/8 Ct R. Biddenham and Newham 8 Apr. 7 Jas; BL, Add. R. 26836; 9288 m. 1; 63591 rental Rickinghall 1579; Add. MS 36875 fo. 29; PRO, St. Ch. Proc. Jas 55/29 m. 2; Sp. Colls, Ct R., G.S. bdl. 195 no. 78 m. [2d.]; Ault, *Husb.*, 26, 43, 49–51, 54, 88–9; *Fmg*, 81sqq., 90sqq., 104sqq., 120–2, 125sqq., 145; 'Vil. By-laws', 378sqq.; 'Early By-laws', 227, 230; 'Man. Ct and Par. Ch. in 15th cent. Engld: stud. vil. by-laws', *Speculum* 1967 xlii, 53sqq., 66–7. For manors: my *Agrarian Probs*, 19–21; Ault, *Fmg*, 121–2. For meetings: Cunningham, op. cit., 208–9; Douglas, op. cit., 160sqq., 231–2, 240, 242; Wake, art. cit., 406, 410–11; W. Cooper, op. cit., app. pp. 27, 35; Vinogradoff, *Growth Man.*, 305–6; *Villainage*, 392–5; Ravensdale, op. cit., 111; Northants. RO, Wmld Coll. 4.xvi.5 Nassington surv. *c.* 1550; PRO, St. Ch. Proc. Jas 55/29 m. 2; Sp. Colls, Ct R., G.S. bdl. 195 no. 78 mm. [2, 3d.]. For sub-committees: PRO, Chanc. Proc. ser. i, Jas B.31/63 complt. For informers: Birm. Ref. Lib. MS 168163, 12 Oct. 1702.

423 W. O. Ault, *Priv. Jurisd. in Engld* Yale Hist. Pubs, Misc. x, New Haven, 1923, pp. 176, 344; 'Vil. By-laws', 389; Homans, op. cit., 319sqq.; G. P. Scrope, *Hist. Man. and Anc. Bny Cas. Combe in Co. Wilts.*, s.l. 1852, p. 337;

H. Hall, *Soc. in Eliz. Age*, 1887, p. 156; J. V. Beckett, *Hist. Laxton: Engld's last open-fld vil.*, Oxford, 1989, p. 34; T. E. Scrutton, *Coms and Com. Flds*, Cambridge, 1887, p. 21; G. Eland, *At Cts Gt Canfield, Ess.*, 1949, pp. 60–1; Northants. RO, Mont. Coll. Northants. old box 17 no. 160 Ct Bk Broughton 12 Oct. 23 Eliz.

424 Ibid., 11 Mar. 1 Mary, 21 Mar. 1 and 2 P.&M.; Aynho Ct R. 9 Apr. 1680; Wmld Coll. 5v.1 Ct R. Woodnewton 15 Sept. 1620; Ches. RO, Nedeham and Kilmorey Coll. Ct Bk Badington and Bromall 22 July 11 Jas; BL, Add. MS 23150 fos 21, 23; PRO, Chanc. Proc. ser. i, Jas B.12/68 m. 3; B.31/63 complt; Ault, *Husb.*, 33, 41–3; *Fmg*, 125–6, 128sqq., 134sqq.; H. M. C. *Middleton MSS*, 109; Crossley, 'Test. Docs', 87.

425 Ibid., 91–2; Dors. RO, D.39/H2 Ct Bk fo. 60; Northants. RO, Mont. Coll. Northants. old box 18 no. 160 Barnwell ords 14 Oct. 28 Eliz.

426 PRO, Chanc. Proc. ser. i, Jas S.29/7 m. 1.

427 PRO, Exch. K. R. Deps by Commn 39 Eliz. East. 26.

428 Wilts. RO, Savernake Coll. Eliz. Ct Bk Collingbourne Ducis, att. sh.; cf. grievances 1584 and sh. inserted (*c.* 1600) pet. to Mr Picke.

429 PRO, Exch. K. R., Deps by Commn 39 Eliz. East. 26; Ault, 'Early By-laws', 228; *Husb.*, 43, 48–9; *Fmg*, 81sqq., esp. 83; my *Agrarian Probs*, 112–13.

430 Wilton Ho. Ct R. Var. Mans 1632 Ct R. Wylye 1632, m. 4 (ptd in Kerridge, *Survs*, 138–40).

431 Ault, *Fmg*, 137, 139, 141–3; *Husb.*, 27–9; my *Ag. Rev.*, 318; Homans, op. cit., 61–3; Bateson et al., op. cit., ix, 119; Oxon Archaeol. Soc. *Rep.* f. 1911 (1912), 136; Shak. Bpl. Man. Docs, ct files Whitchurch, ords 4 Oct. 1582; Ct R. Henley-in-Arden 22 Oct. 25 Chas 2; W. de B. Coll. 703b, 12 Apr. 1616; 1194, 6 Apr. 11 Chas; PRO, D.L., Ct R. 81/119, m. 14d.; Beds. RO, X.69/6 Goldington pains 1687; Seebohm, op. cit., 450; Eland, *Gt Canfield*, 16, 90.

432 Plot, *Oxon.*, 152; T. Nourse, *Campania Foelix*, 1700, p. 27; my *Ag. Rev.*, 181, 195sqq.; *Agrarian Probs*, 119.

433 Ibid., 108–9; H. S. Darbyshire and G. D. Lumb, *Hist. Methley* Thoresby Soc. 1937 (1934) xxxv, 65–6; F. R. Twemlow, 'Man. Tyrley in Co. Stafford', *Colls Hist. Staffs.* 3rd ser. 1948 (1945–6), 140; Crossley, 'Test. Docs', 88; Brookes, op. cit., 132; Hoskins, 'Flds Wigston Magna', 173–4; Ault, *Husb.*, 93; Boyd, op. cit., 25–6; Stats. 24 Hy 8 c.10; 8 Eliz. c.15; S. C. Ratcliffe and H. C. Johnson, *War. Co. Recs iv: Q. Sess. Ord. Bk*, Warwick, 1938, p. 76; *Trans. Cambs. and Hunts. Archaeol. Soc.* 1904 i, 153, 160; Wilton Ho. Ct R. 1666–89, Stoke Verdon 25 Nov. 1667; Bucks. RO, 10/48 Ct R. Long Crendon 15 May 42 Eliz.; Lancs. RO, DD. Pt. 22 Ct Bk 1–8 Eliz. p. 6; War. RO, M.R. 5 Moreton Morrell 9 Oct. 1710; Leics. RO, Beaumanor Coll. Ct files Barrow-on-Soar 15 Aug. 1723; Shak. Bpl. Man. Docs, Ct R. Alveston and Tiddington 11 Apr. 18 Chas 2; Leigh Coll. Ct R. Stoneleigh 5 Oct. 23 Eliz.; BL, Add. MS 36875 fo. 25; PRO, D.L., Ct R. 84/1155 Ct files Pailton 10 Nov. 1801, 3 Nov. 1813; Northants. RO, Daventry Coll. 540 Ct R. Hellidon 1660; I(L) Coll. 128 Ct R. Lamport 6 Apr. 11 Eliz.; 3545 fo. 35; Old Par. Recs, Fld Offrs' Acct Bk 1744; Mont. Coll. Northants. old box 7 no. 66/5 Ct files Weekley 21 Mar. 9 Eliz.; no. 72 Ct Bk id. 3 Apr. 15 Eliz.; box 17 no. 160 Ct Bk Broughton 27 Apr. 21 Eliz.; box 18 no. 160 Ct R. Barnwell 1 Dec. 1614, 9 Dec. 5 Chas;

War. old box 6 no. 14 (SR65) Ct R. Newbold 17 Oct. 1699, 27 Oct. 1707; Misc. Led. 145 p. 543.

434 Ibid., Northants. old box 17 no. 160 Ct Bk Broughton 30 Mar. 15 Eliz.; Wimbledon Com. Cmmttee, *Exts fr. Ct R. Man. Wimbledon*, 1866, p. 93; C. T. Clay, *Yks Deeds v* Yks. Archaeol. Soc. Rec. Ser. 1926 lxix, 99; Bateson et al., op. cit., ix, 194; Seebohm, op. cit., 449–50; Wilton Ho. Ct R. Mans 1689–1723 box i, vol. iv, p. 60; box ii, vol. ii, pp. 43–4, 50, 57; Ches. RO, Vernon (Warren) Coll. Ct R. nos 7, 9; Bucks. RO, 616/43 Ct R. Emberton 9 Apr. 15 Chas 2; Shak. Bpl. Leigh Coll. Ct R. Adlestrop 24 Oct. 1704, 20 Mar. 1758; Deene Ho. Brudenell MS O.xxiv.1 Ct Bk Stonton Wyville 28 Apr. 16 Eliz., 4 Dec. 19 Eliz.; Ault, *Fmg*, 126, 138, 140–1, 143, 146; PRO, Exch. K.R., Deps by Commn 16–17 Eliz. Mich. 9; 17 Eliz. East. 1; Webb, op. cit., pt i, p. 87.

435 Ibid.; Bucks. RO, 155/21 Ct files Taplow 28 Sept. 1620; Dors. RO, D.4 Thornhull Ct Bk sect. 1, 17 Oct. 1544, 6 Nov. 4 Ed. 6; Leics. RO, Beaumanor Coll., Barrow Ct R., Quorndon pains 8 Oct. 1571; PRO, D.L., Ct R. bdl. 106 no. 1529 m. 4; Northants. RO, Wmld Coll. 5.v.1, Ct R. Stanground 16 Sept. 22 Jas; Mont. Coll. Bucks. box P, pt 2, X.890, Ct files 16 Apr. 36 Chas 2; Northants. old box 7 no. 72 Ct Bk Weekley 18 May 8 Eliz., 11 Apr. 10 Eliz., 16 Oct. 18 Eliz.; box 18 no. 160 Ct R. Barnwell 1 Dec. 1614; box 25 no. 52 Ct Bk Luddington 23 Oct. 20 Eliz.; Old Par. Recs, Fld Offrs' Acct Bk 1739–44, 1752; Aynho Ct R. 20 Oct. 1 Jas 2.

436 Ibid., Ecton Coll. 1191, art. 24; Daventry Coll. 549b, arts 5–7; Webb, loc. cit.; my *Ag. Rev.*, 291–3; Barley, art. cit., 51; Shak. Bpl. Man. Docs, Ct R. Alveston and Tiddington 30 Oct. 1713, 25 Oct. 1719, 21 Oct. 1720, 30 Oct. 1724, 13 Oct. 1730, 20 Oct. 1732.

437 Ibid., Ct R. Fenny Drayton 17 Oct. 1683; Ct R. Claverdon 1515–1779, pass.; Leigh Coll. Fillingley pains and ords 12 Oct. 1704; Stark, op. cit., 91; Marshall, *Yks.*, i, 28–9; Bateson et al., op. cit., vii, 316; Lancs. RO, DD.X/80/1; BL, Add. R. 6284 mm. d, e, f; 26885–91; 28258; Bucks. RO, W10/1 W. Wycombe Ct R., 27 Sept. 19 Jas; Herts. RO, Ashridge Coll. nos 160–7 Ct R. Southall in Gt Gaddesden; Cashiobury and Gape Coll. 6387, 6419; Soms. RO, DD/SAS Acc. C.432, Ct R. Old Cleeve 7 Oct. 9 Jas, 7 Oct. 10 Jas, et pass.; Northants. RO, Misc. Led. 145, pp. 27–8; PRO, Sp. Colls, Ct R., G.S. 194/59 m. 2; 195/82 m. 1d.; D.L., Ct R. 81/1114 mm. 2, 4, 5; 1115 mm. 2, 4, 6, 8; 1120; 82/1132 m. 2; 84/1155, Ct files Brinklow 10 Nov. 1801.

438 Ault, *Husb.*, 87.

439 Leconfield, *Petworth*, 36, 39; Birm. Ref. Lib. 505455; BL, Add. MS 38487 fo. 12.

440 Northants. RO, Mont. Coll. Hunts. old box 26 no. 28 Ct R. Bk Caldecote 2 Apr. 14 Eliz.; PRO, Exch. L.R., M.B. 221 fo. 19; cf. K.R., Deps by Commn 22 Jas Hil. 13; A.O., Parl. Survs: Corn. 25 fos 1, 2; Worcs. 7 fo. 3.

441 Ibid., Notts. 13 fo. 2; Sp. Colls, Ct R., G.S. 195/36 m. 2d.; 37 m. 1d.; 82 m. 1d.; D.L., Ct R. 82/1131 m. 2; 105/1504 m. 2; Beds. RO, L.26/10 Ct R. Bedlow 13 Oct. 1686; Herts. RO, Broxborne Bury Coll. 58 p. 113; Sfk RO (Ips.) V.5.18.10.1 p. 19; Lancs. RO, DD. F. 158; Scrope, op. cit., 336–7; Wake, art. cit., 412; Ault, *Fmg*, 117, 129–30, 142; *Husb.*, 69, 73, 78, 81.

442 Birm. Ref. Lib. 505455; BL, Add. MS 36908 fo. 8v; Deene Ho. Brudenell MS O.xxiv.1 Ct Bk Hougham 16 Oct. 35 Eliz.; Deene Oct. 18 Eliz.; Thistleton 8 Apr. 34 Eliz.; Stonton Wyville 15 Apr. 33 Eliz.

443 Ibid., I.v.8 Glapthorne 25 Sept. 20 Eliz.; Leics. RO, Beaumanor Coll. Ct R. Beaumanor 22 Oct. 1723.

444 Hull, op. cit., 605, 608; Northants. RO, Mont. Coll. Northants. old box 16 no. 98 Ct Bk L. Oakley 20 May 3 and 4 P.&M.; Wmld Coll. 5.iii.1 Ct R. Woodston 17 Sept. 1621; 5.v.1 Ct R. Stanground 18 Sept. 10 Jas; Aynho Ct R. Sept. 18 Jas; Deene Ho. Brudenell MSS, I.v.55 Ct file Glapthorne 12 Apr. 1610; I.v.62a ords 1608; Wilts. RO, Keevil and Bulkington Ct Bk 44 Eliz. – 2 Chas, 29 Sept. 21 Jas; War. RO, M.R. 21 no. 2 Ct R. Berkswell 10 Apr. 1616; PRO, D.L., Ct R. bdl. 80 no. 1101 Cas. Donington 7 Oct. 6 Eliz.; Sp. Colls, Ct R. G.S. bdl. 195 no. 36 m. 2d.; Shak. Bpl. Man. Docs, Ct Bk Alveston and Tiddington 22 Oct. 8 Jas; Lancs. RO, DD.F.158; BL, Harl. MS 702 fo. 14; Staffs. RO, Hand Morgan Coll. Chetwynd MSS, Ct R. Shenstone 1 Oct. 7 Chas; Ault, *Ct R. Ramsey*, 174, 185, 196, 199, 204, 207, 210–11, 213, 228–9; *Fmg*, 129, 141; *Husb.*, 89, 90, 92.

445 Ibid., 19, 20, 91–2; Shak. Bpl. W. de B. Coll. 703a Exts Ct R. Barford 10 Oct. 1605; Bucks. RO, 452/39 Ct R. Stone, 30 Sept. 12 Jas; BL, Add. MS 23150 fo. 16; Leics. RO, Ct files Belton 26 Oct. 1681; Northants. RO, I (L) Coll. 128 Ct R. Lamport 6 Apr. 11 Eliz.; PRO, Req. Eliz. 129/45; my *Ag. Rev.*, 93.

446 Baigent and Millard, op. cit., 334; Ault, *Fmg*, 120, 139; Northants. RO, Wmld Coll. 5.v.1 Ct R. Farcet 13 Sept. 1623, 17 Sept. 1624; Mont. Coll. Northants. old box 18 no. 160 Barnwell ords 19 Oct. 1613, 9 Dec. 5 Chas.

447 Ibid., old box 25 no. 51 Ct R. Luddington 23 Apr. 15 Eliz.; Whiteman, op. cit., 112; PRO, Exch. K.R. Sp. Commn 4684; Sp. Colls, Ct R. G.S., 194/54 Stanwick 8 June 2 Eliz.

448 Baigent and Millard, op. cit., 350; *Wimbledon*, 95; Lancs. RO, DD. K.1526/5, 17 Apr. 6 Geo. 2; PRO, Exch. L.R., M.B. 207 fo. 30 (28); 214 fo. 82; A.O., Parl. Survs Derbys. 19 fo. 21.

449 Ibid., Ches. 19 fo. 11; BL, Add. R. 9283 m. 1; 9288 m. 1.

450 PRO, St. Ch. Proc. Jas 40/22 m. 110; Ault, *Husb.*, 54, 89.

451 Ibid., 90, 92–3; Barley, art. cit., 51; Raistrick, op. cit., 46; Bateson et al., op. cit., viii, 242; Cornwall, 'Fmg', 52–3; E. Cannan, *Hist. Loc. Rates in Engld*, 1896, pp. 10–12, 18–20, 22–5, 42, 50, 102; Sfk RO (Ips.) V.5.18.10.1 p. 19; Shak. Bpl. Leigh Coll. Ct R. Stoneleigh 5 Oct. 23 Eliz.; BL, Add. MS 36585 fo. 163; Add. R. 26836 Ct R. Glatton cum Holme 1 Oct. 33 Eliz.; Berks. RO, D/EPb M3 Ct Bk Coleshill 1592–1687 p. 66; Bucks. RO, 10/48 Ct R. Long Crendon 15 May 42 Eliz.; W. Cooper, op. cit., app. p. 35; Lancs. RO, DD.K.1506/2 Ct R. Burscough 27 Oct. 29 Chas 2; Northants. RO, I (L) Coll. 128 Ct R. Lamport 2 May 29 Eliz.; YZ Coll. 988 Ct R. Greatworth 12 Jan. 1628; Aynho Ct R. 19 Apr. 19 Chas 2; Wmld Coll. 5.v.1 Ct R. Apethorpe 18 Sept. 9 Jas; Ches. RO, Vernon (Warren) Coll. Ct R. no. 5.

452 Dors. RO, Weld Coll. D.10/E.103, 'Orders 1665', no. 10.

453 BL, Harl. MS 6006 fo. 17; Northants. RO, Daventry Coll. D.541 Ct R. Hellidon 15 Oct. 1669.

454 Ibid., Mont. Coll. Northants. old box 18 no. 160 Barnwell ords 1 Dec. 1614, 25 May 1613; Misc. Led. 145 pp. 183, 535; 146 p. 12; Ecton Coll. 1183 Ct R. Ecton 9 Oct. 1713; Wmld Coll. 5.v.1 Ct R. Woodston 12 Sept. 1623; Farcet 17 Sept. 1624, 25 Oct. 1660; Wake, art. cit., 410; Ault, *Husb.*, 44–6, 49, 90; S.C. Powell, op. cit., 28, 151, 173–5, 184–5; Crossley, loc. cit., 89, 90, 101–2; Jeffery, op. cit., 24; BL, Harl. MS 702 fo. 14; Add. MS 36585 fo. 164; PRO, Sp. Colls, Ct R., G.S. 195/37 m. 4; D.L., Ct R. 80/1101 Cas. Donington 7 Oct. 6 Eliz.; 106/1529 m. 1d.; Req. Eliz. 129/45; Ches. RO, Vernon (Warren) Coll. Ct R. nos 7, 15; Shak. Bpl. W. de B. Coll. 1194; Trevelyan Coll. Ct R. Snitterfield and Bearley 27 Oct. 1713; Leics. RO, DE.40.40 Ct Bk Ullesthorpe 1 Oct. 1690; Ct file Belton, pains 26 Oct. 1681.

455 Blith, op. cit., 75; E. Laurence, op. cit., 82–3; PRO, Chanc. Proc. ser. i, Jas C.7/14 complt; Wilts. RO, Savernake Coll. Eliz. Ct Bk Collingbourne Ducis, att. sh.

456 PRO, Req. Eliz. bdl. 129 no. 45 answ.

457 PRO, Exch. Treas. Recpt, Bk 157 fo. 64 (p. 121); cf. Wilts. RO, Acc. 212B, B.H.8, Broad Hinton rental 1636.

458 PRO, Exch. L.R., M.B. 221 fo. 325; Ess. RO, D/DL.M.81 bks 1, 2.

459 Northants. RO, F–H Coll. 1624 Ct R. Church Brampton, att. n. ords; and as n. 407.

460 Wilts. RO, Acc. 212B Ct R. Grittleton 1 Oct. 6 Chas; Cambs. RO, L.19.17 Ct R. Haslingfield 30 Apr. 1655; BL, Add. MS 36920 fo. 83v.

461 G. Goodman, *Ct K. Jas I* ed. J.S. Brewer, 2 vols, 1839, i, 342.

462 Marshall, *Yks.*, i, 28–9; *Glos.*, ii, 13; *Midl. Dept*, 215; *W. Engld*, i, 21–2.

463 Ibid., 22.

464 J. Aubrey, *Wilts. Topog. Colls 1659–70* ed. J.E. Jackson, Devizes, 1862, p. 9; 'Intro. to Surv. and Nat. Hist. N. Div. Wilts.' in *Miscs on Sev. Curious Subjs*, E. Curll, 1714, p. 30.

465 H.L., *MSS H.L.* n.s., sev. vols, 1900– in prog. xi, Addenda 1514–1714 (ed. M.F. Bond) 1962, pp. 3–5; Stats 24 Hy 8 c. 10; 8 Eliz. c. 15; 13 Geo. 3 c. 81; G.R. Elton, *Parl. Engld 1559–81*, Cambridge, 1986, p. 235; [R. Bancroft], *Dangerous Posns and Proc.*, 1593, pp. 45, 47; Gonner, op. cit., 57; Ashton, op. cit., 60; Notestein et al., op. cit., ii, 173.

466 M.K. Ashby, *Changing Engl. Vil.: hist. Bledington, Glos. in its setting, 1066–1914*, Kineton, 1974, pp. 137, 141–4; Boyd, op. cit., 79; A.C. Chibnall, op. cit., 286–9; Leics. RO, Leire arts agrt 1689; Bucks. RO, Swanbourne indre agrt 7 Nov. 22 Geo. 2; Acq. 35/39, Bye-laws and Regs f. Padbury Open Flds 1779; Hoskins, 'Flds Wigston Magna', 174–5; Cn Bennett, art. cit., 36sqq.; Glouc. Ref. Lib. MS RF.30/3; Northants. RO, Daventry Coll. 573a Ravensthorpe agrt f. com. ords 1721.

467 Ibid., 549b Hellidon ords 27 Oct. 1740 arts 5, 6; Ecton Coll. 1189 Ct R. Ecton 17 May 1701; Old Par. Recs, Fld offrs' accts 1738–9, 1744, 1752; Leics. RO, Beaumanor Coll. Ct R. and files Barrow etc. Quorndon 26 Oct. 1756; BL, Add. R. 32392; Bushby and Le Hardy, op. cit., 125; Crossley, 'Test. Docs', 89.

468 Stats 22 Hy 8 cc.5, 12; 24 Hy 8 c.10; 27 Hy 8 c.25; 2 and 3 P.&M. c.8; 8 Eliz. c.15; 18 Eliz. c.3; 39 Eliz. c.3; D. H. Willson, *Parl. Diary Robt Bowyer*

1606–7, Minneapolis, 1931, p. 35; E. J. King, *Yrs Beyond Mem.*, Oxford, 1954, pp. 28–9, 45.

469 Ibid., 29, 45; G. A. Thornton, 'Stud. in Hist. Clare, Sfk', *Trans. R. Hist. S.* 4th ser. 1928 xi, 110–11; W. E. Tate, *Par. Chest*, Cambridge, 1951, pp. 256–7.

470 Tawney, op. cit., 14.

471 Plot, *Oxon.*, 152; Pitt, *Leics.*, 80; Nourse, op. cit., 27.

472 Bateson et al., op. cit., ii, 417; iv, 139–40; Marshall, *Midl. Dept*, 466; Ault, *Husb.*, 36; J. R. Ravensdale, 'Landbeach in 1549: Ket's Rebellion in Miniature' in L. M. Mundy, *E. A. Studs*, Cambridge, 1968, p. 111; Northants. RO, Wmld Coll. 4.xvi.5 Nassington surv. *c.* 1550; Deene Ho. Brudenell MS B.ii.1; BL, Egerton MS 3007 fo. 130; Glos. RO, D.445/M12 (1534); PRO, Exch. L.R., M.B. 221 fo. 325; K.R. Deps by Commn 17 Eliz. East. 1; Chanc. Proc. ser. i, Jas H.14/34; H.36/74 mm. 1sqq., 9; L.13/63; S.4/41; Neilson, *Econ. Conds*, 62–3; 'Proc. in St. Ch. Hy VIII and Ed. VI', *Colls Hist. Staffs.* 1912, pp. 69, 70; Thirsk, 'Com. Flds', 25.

473 Marshall, *S. Cos*, ii, 319, 350–1.

474 Ernle, Ld, *Engl. Fmg Past and Pres.*, 1927, pp. 59, 65.

475 My *Ag. Rev.*, 135–6.

476 Bateson et al., op. cit., ii, 418; xii, 110.

477 James and Malcolm, *Bucks.*, 29; Vancouver, *Cambs.*, 148; Tawney, op. cit., 220–3.

478 Bateson et al., op. cit., ii, 418; Elliott, 'Aspatria', 100; G. F. Farnham, *Quorndon Recs*, 1912, p. 14; Leics. RO, Beaumanor Coll. Ct R. and files Barrow etc. Woodhouse ords 1560.

479 My *Ag. Rev.*, 290.

480 Marshall, *Yks.*, i, 51; R. de Z. Hall, 'Post-Med. Ld Tenure, Preston Plucknett', *Rep. Soms. Archaeol. and Nat. Hist. Soc.*, 1961 cv, 114. Point about risk is unduly laboured in D. N. McCloskey, 'Engl. Open Flds as Behavior towards Risk' in P. Uselding, *Res. in Econ. Hist.: ann. comp.* 1976 i; 'Persistence Engl. Com. Flds' in W. N. Parker and E. L. Jones, *Eur. Peasants and their Mkts*, Princeton, N.J., 1975; cf. C. J. Dahlman, *Open-Fld Syst. and Beyond*, Cambridge, 1980, pp. 58sqq.

481 Tusser, op. cit., 134sqq., 207; Fitzherbert, *Surveyinge* fos 52–3; my *Ag. Rev.*, 202sqq.

482 E.g. Pitt, *Northants.*, 38; Vancouver, *Ess.*, 105, 107–8.

483 Marshall, *Glos.*, ii, 10; *Yks.*, i, 292; S. Fortrey, *Engld's Int. and Improvemt*, Cambridge, 1663, pp. 15, 16, 18.

484 Marshall, *Midl. Cos*, i, 18; Fitzherbert, *Surveyinge* fo. 52; Pitt, *Northants.*, 33–4, 36, 38; Barratt, op. cit., i, 128; I. Leadam, *Dom. Inclos*, 2 vols, 1897, i, 66.

485 My *Agrarian Probs*, 128–30, 132–3; my *Ag. Rev.*, 209.

486 Marshall, *Glos.*, i, 21; *S. Dept*, 165; Fitzherbert, *Surveyinge* fo. 52; Norden, *Surveiors Dial.*, 99; Blith, op. cit., 74; T. Davis, op. cit. (1813), 44–5; J. Wedge, *Gen. View Ag. Co. War.*, 1794, pp. 20–1; cf. Yelling, op. cit., 211–12.

487 W. Marshall, *On Landed Prop. Engld*, 1804, p. 13.

488 T. Davis, op. cit. (1794), 84.

489 Bloch, op. cit., 26–9, 35, 48–9, 51–2, 60–1; Haudricourt and Delamarre, op. cit., 61sqq., 92sqq., 109–11, 139, 204sqq., 224sqq., 253sqq.,

276sqq., 287sqq., 378–9; Pliny, op. cit., lib. xviii, cap. xlviii; A. Dopsch, *Econ. and Soc. Founds Eur. Civ.*, 1937, p. 162; D. Warriner, *Contrasts in Emerging Socs*, 1965, pp. 51, 143–4, 217, 309, 311, 323; F. Braudel, *Civ. and Capitalism 15th to 18th cent.*, 3 vols, 1981–4, i, 118; J. H. Clapham, *Econ. Dev. Fr. and Ger. 1815–1914*, Cambridge, 1936, pp. 7, 30; L. von Thallóczy, *Illyrisch–Albanische Forschungen*, 2 vols, Munich and Leipzig, 1916 ii, 14–16; S. Ilešič, *Die Flurformen Sloweniens im Lichte der europäischen Flurforschung* (Münchner Geogr. Hefte xvi 1959, pp. 75, 78; M. Defourneaux, 'Le Problème de la terre en Andalousie au XVIIIe s. et les projets de réforme agraire', *Revue Hist.* 1957 ccxvii, 48; J. Weulersse, *Paysans de Syrie et du Proche-Orient*, Paris, 1946, pp. 99, 153.

490 Ibid., 104–5, 107, 145sqq., 151–3; Dopsch, op. cit., 141–2, 162; D. Warriner, *Ld Ref. and Dev. in M. E.*, 1962, pp. 43–4, 98; *Contrasts*, 307–9, 322; J. Day, 'Peuplement, cultures et régimes fonciers en Trexenta (Sardaigne) XIIIe–XIXe s.' in A. Guarducci, *Agricoltura e Trasformazione dell'Ambiente secoli XIII–XVIII*, Prato, 1984, pp. 699, 700; F. Irsigler, 'Intensivwirtschaft, Sonderkulturen u. Gartenbau als Elemente der Kulturlandschaftgestaltung in den Rheinlanden (13.–16. Jh.)', ibid., 729sqq.; C. Vanzetti, 'L'Utilisation du Sol', ibid., 114–15; V. Zimányi, 'La Formation des monocultures et ses conséquences écologiques en Hongrie', ibid., 57–9, 64; A. Meyer, 'La Structure agraire dans la commune d'Otterwiller', *Ann. de Géog.* 1937 xlvi, 127, 131–2; R. Livet, 'Les Champs allongés de Basse-Provence' in *Géog. et Hist. Agraires* Ann. de l'Est Mém. 21, pp. 383sqq.; X. de Planhol, 'Essai sur la genèse du paysage rural de champs ouverts', ibid., 416, 421; W. Müller-Wille, 'Langstreifenflur u. Drubbel: ein Beitr. z. Siedlungsgeogr. Westgermaniens', *Dtsch. Archiv f. Landes- u. Volksforschung* 1944 viii, 36; E. Juillard, *La Vie rurale dans la plaine de Basse-Alsace*, Paris, 1953, pp. 33, 37, 41, 43, 46, 263; 'L'Assolement biennal dans l'agriculture septentrionale: le cas particulier de la Basse-Alsace', *Ann. de Géog.* 1952 li, f.p. 34, pp. 35–7, 43, 45; Clapham, op. cit., 8, 32–3; G. Hanssen, *Agrarhistorische Abh.*, 2 vols, Leipzig, 1880–4, i, 166; E. Laur, *Swiss Fmg*, Berne, 1949, pp. 36, 64; C. Jireček, *Gesch. der Serben*, 2 vols, Gotha, 1911–18, ii, 54; I. Balassa, 'Der Maisbau in Ungarn', *Acta Ethnographica Academiae Scientiarum Hungaricae* 1956 v, fasc. 1–2, pp. 111–12; M. Belényesy, 'Angaben über die Verbreitung der Zwei- u. Dreifelderwirtschaft im Mittelalterlichen Ungarn', ibid., 183sqq.; H. Sée, *Esquisse d'une hist. du régime agraire en Eur. aux XVIIIe et XIXe s.*, Paris, 1921, pp. 38–9; G. E. Fussell, 'Fmg Systs in Cl. Era', *Tech. and Culture* 1967 viii, 23–4, 29, 31–2, 38, 41; A. R. Lewis, *N. Seas: shipg and comm. in N. Eur. 300–1100*, New York, 1978, pp. 29, 90–1, 199; Braudel, *Civ.*, i, 114; *Ident. Fr.*, 2 vols, 1988–90, ii, 267, 269; Schröder-Lembke, 'Wesen u. Verbreitung der Zweifelderwirtschaft im Rheingebiet', *Zs. f. Agrargesch. u. Agrarsoziologie* 1955 vii, 14sqq., 26; 'Römische Dreifelderwirtschaft?', ibid., 1963 xi, 25sqq. is much best treatment of subject; E. Klag, 'Ein Beitr. z. Zweifelderwirtschaft', ibid., 1969 xvii, 52, 55; A. Mayhew, *Rural Set. and Fmg in Ger.*, 1973, pp. 85, 170; Ilešič, *Flurformen*, 78–9, 82; R. Krzymowski, *Die landwirtschaftlichen Wirtschaftssysteme Elsass-Lothringens*, Gebweiler im Elsass, 1914, pp. 190sqq., 199sqq., 214sqq., 231sqq., 237, 239, 244–6, 333sqq., 337, 339, 341sqq., end map; W. Abel, *Gesch. der dtsch. Landw. vom frühen Mittelalter bis zum 19. Jh.*, Stuttgart, 1967, pp. 87, 217, 220; M. Bloch, *Fr. Rural Hist.: essay on its basic chars*, 1966, pp. 30sqq., 48sqq.; Thallóczy, op. cit., 15; K. Lamprecht, *Dtsch. Wirtschaftsleben im Mittelalter*, vol. i, pt i, Leipzig, 1886, p. 545; Pliny, op. cit., lib. xviii, cap. x; Virgil, *Georgics* bk i, ll. 47–8.

491 O. A. Johnsen, *Norwegische Wirtschaftsgesch.*, Jena, 1939, p. 408; E. Heckscher, *Econ. Hist. Swed.*, Cambridge, Mass., 1963, pp. 26, 28–9, 115–16, 153–4; K. Wührer, *Beitr. z. ältesten Agrargesch. des germanischen Nordens*, Jena, 1935, p. 98; A. Meitzen, *Siedelung u. Agrarwesen der Westgermanen u. Ostgermanen, der Kelten, Römer, Finnen u. Slawen*, 3 vols, Berlin, 1895, i, 69, 70.

492 Braudel, *Civ.*, i, 147; *Rep. R. Cmmn Ag. in Ind.*, 1928, p. 5; B. H. Baden-Powell, *Ld Systs Brit. Ind.*, 3 vols, Oxford, 1892, i, 116, 119; *Orig. and Growth Vil. Communs in Ind.*, 1899, pp. 104–5, 120; *Ind. Vil. Commun.*, 1896, pp. 50, 52–3, 256–7.

493 Ibid., 271; G. Keating, *Rural Econ. in Bom. Deccan*, 1912, pp. 41, 45, 111; G. Slater, *Econ. Studs vol. i: some S. Ind. Vils*, London, Bombay, Madras and New York 1918, pp. 19, 20, 79, 115, 208, 243.

494 *Outlines Ag. in Jap.*, Dept Ag. and Comm., Tokyo, 1910, p. 22; T. C. Smith, *Agrarian Origs Mod. Jap.*, Stanford, Cal., 1959, pp. 12, 19; Hsiao-Tung Fei, *Peasant Life in Ch.*, 1939, pp. 195–6; A. Donnithorne, *Ch's Econ. Syst.*, 1967, p. 31.

495 Baden-Powell, *Ld Systs*, i, 119; iii, 490, 508; J. S. Lewinski, *Orig. Prop. and Formn Vil. Commun.*, 1913, pp. 13–16; Braudel, *Civ.*, i, 147.

496 J. Blum, *Noble Landowners and Ag. in Aus. 1815–48: stud. in origs peasant emancipation*, Baltimore, 1947, p. 157; *Ru.*, 21, 166, 231, 337, 515; Abel, op. cit., 82–5, 217; Krzymowski, op. cit., 94sqq., 119sqq., 129sqq., 145sqq., 154sqq., 161sqq., 175–8, end map; J. Mavor, *Econ. Hist. Ru.*, 2 vols, 1925, ii, 274–5, 286; G. Hard, 'Ein Vierzelgensystem des 18. Jh', *Zs. f. Agrargesch. u. Agrarsoziologie* 1963 xi, 160; Mayhew, op. cit., 86, 170; A. Hömberg, *Siedlungsgesch. des oberen Sauerlandes* Geschl. Arb. z. westf. Landesforschung, Bd 3, Münster, 1938, pp. 93–4; P. I. Lyaschenko, *Hist. Nat. Econ. Ru. to 1917 Rev.*, New York, 1949, pp. 179–80; M. Kovalevsky, 'L'Agriculture en Russie' rep. fr. *Revue Internationale de Sociologie*, Paris, 1897, p. 6; L. Żytkowicz, 'Peasant's Fm and Landlord's Fm in Pol. fr. 16th to mid. 18th cent.', *Jnl Eur. Econ. Hist.* 1972 i, 138, 146; L. Musset, 'Observations sur l'ancien assolement biennal du Roumois et du Lieuvin', *Ann. de Normandie* 1952 ii, 143; C. Parain, 'Travaux récents sur l'hist. rurale de Danemark', ibid., 127; Belényesy, art. cit., 183, 188; K. H. Schröder, 'Die Flurformen Württemberg-Hohenzollerns u. ihre neuzeitliche Umgestaltung', *Raumforschung u. Raumordnung*, 1941, v, 310–11, 320; A. Krenzlin, 'Probleme der neueren norddeutschen u. ostmitteldeutschen Flurformenforschung', *Dtsch. Archiv f. Landes- u. Volksforschung* 1940 iv, 556, 566–7; W. Müller-Wille, 'Das Rheinische Schiefergebirge u. seine kulturgeogr. Struktur u. Stellung', ibid., 1942 vi, 537, 557sqq.; *Die Ackerfluren im Landesteil Birkenfeld u. ihre Wandlungen seit dem 17. u. 18. Jh.*, Bonn, 1936, pp. 12–14, 23, 35, 56sqq., 86, 96sqq.; 'Langstreifenflur', 28–30, 36; A. Krenzlin and L. Reusch, *Die Entstehung der Gewannflur nach Untersuchungen im nördlichen Unterfranken*, Frankfurt a.M., 1961, p. 115; Hanssen, op. cit., i, 127–9, 132sqq., 139–40, 190sqq.; F. Steinbach u. E. Becker, *Geschl. Grundlagen der kommunalen Selbstverwaltung in Dtschl.* Rhein. Archiv xx, Bonn, 1932, p. 42; J. De Vries, *Du. Rural Econ. in Golden Age 1500–1700*, New Haven and London, 1974, pp. 33,

241; J. Bieleman, *Boeren op het Drentse Zand 1600–1910: een nieuwe visie op het 'oude' landbouw* (*A. A. G. Bijdragen*, vol. xxix), Wageningen, 1987, pp. 531, 591sqq.; V. Dorošenko, 'Der ostbaltische Herrenhof des 16.–18. Jh. als "Getreidefabrik" ' in Guarducci, op. cit., 208; *Die Landw. in Bayern*, Munich, 1890, p. 155; Sée, op. cit., 157; my *Ag. Rev.*, 151; Clapham, op. cit., 30.

497 Ibid.; Heckscher, op. cit., 26; Baden-Powell, *Ld Systs*, ii, 679; iii, 490; *Orig.*, 95, 104–5, 120; *Ind. Vil. Commun.*, 257.

498 Blum, *Aus.*, 157–8; *Landw. in Bayern*, 152; Johnsen, op. cit., 408; Abel, op. cit., 219; Mayhew, op. cit., 86; A. Nielsen, *Dänische Wirtschaftsgesch.*, Jena, 1933, pp. 181–2; Hanssen, op. cit., i, 132sqq., 216sqq.

499 De Vries, op. cit., 34; B. H. Slicher van Bath, 'Econ. and Soc. Conds in Frisian Dists fr. 900 to 1500', *A. A. G. Bijdragen* 1965 xiii, 109; A. E. Verhulst, 'Probleme der Mittelalterlichen Agrarlandschaft in Flandern', *Zs. f. Agrargesch. u. Agrarsoziologie* 1961 ix, 14, 19.

500 Baden-Powell, *Ld Systs*, i, 159–61; ii, 136–7, 141–2, 679; *Ind. Vil. Commun.*, 414; Ilešič, *Flurformen*, bet. 16–17, 24, 27, 41, 74, 76, 79, 90–3, 95, 97sqq., 123–4; 'Die jüngeren Gewannfluren in Nordwestjugoslawien', *Geografiska Annaler* 1961 xliii, 130–1, 133, 135; A. Krenzlin, 'Z. Genese der Gewannflur in Dtschl. nach Untersuchungen im nördlichen Unterfranken', ibid., 200; 'Probleme', 547sqq., 556, 561, 567; Schröder, art. cit., 310sqq., 317–18, 320; Krenzlin and Reusch, op. cit., 79, 94, 96, 110; Müller-Wille, 'Langstreifenflur', 19, 21, 23–4, 28–9, 34, 36, 38, 40; F. Steinbach, 'Geschl. Siedlungsformen in der Rheinprovinz: Aufgaben der Siedlungsgesch. in der Rheinprovinz', *Zs. des Rhein. Ver.* 1937 xxx, 24–5; Bloch, op. cit., 36.

501 Ibid., 40–1, pls iv, v; Parain, art. cit., 127; G. W. Coopland, *Abb. St Bertin and its Neighbourhood 900–1350* Oxf. Studs in Soc. and Leg. Hist. iv, Oxford, 1914, pp. 122–3.

502 W. H. Bruford, *Ger. in 18th cent.: soc. backgd lit. revival*, Cambridge, 1959, p. 114; Ilešič, *Flurformen*, 52–4, 122; Lamprecht, op. cit., I, i, 545; G. von Below, *Gesch. der dtsch. Landw. des Mittelalters in ihren Grundzügen*, Jena, 1937, p. 42; H. and G. Mortensen, *Die Besiedlung des nordöstlichen Ostpreussens bis zum Beginn des 17. Jh.*, 2 pts, 'Dtschl. u. der Osten' ser., vols 7, 8, Leipzig, 1937–8, i, 97; ii, 43; Abel, op. cit., 18.

503 My *Ag. Rev.*, 159–60; as nn. 217–221.

504 A. V. Chayanov, *Theory Peasant Econ.* ed. D. Thorner, B. Kerblay and R. E. F. Smith, Homewood, Ill., 1966, p. 141; cf. Lyaschenko, op. cit., 179–80.

505 Müller-Wille, *Birkenfeld*, 58–9.

506 Bieleman, op. cit., 529–31, 539, 574–5, 579–81, 591sqq.; Bloch, op. cit., 32–3; Braudel, *Ident.*, ii, [353].

507 Weulersse, op. cit., 153.

508 Johnsen, op. cit., 408.

509 Krenzlin, 'Probleme', 547 sqq., 556–7; Müller-Wille, 'Langstreifenflur', 28–9; Ilešič, *Flurformen*, 52–4, 73, 75; 'Les Problèmes du paysage rural en Yougoslavie nord-occidentale et spécialement en Slovénie' in *Géog. et Hist. Agraires* Ann. de l'Est Mém. 21, p. 285; A. Krenzlin, 'Blockflur, Langstreifenflur u. Gewannflur als Ausdruck agrarischer Wirtschaftsformen in Dtschl.', ibid., 353sqq.

510 Ibid., 353–4, 364; Bieleman, op. cit., 531, 539, 574sqq., 591sqq., 604sqq.; Parain, art. cit., 127; Abel, op. cit., 85–6, 217–18; Mayhew, op. cit., 85, 169; Hanssen, op. cit., i, 192–3; cf. B. H. Slicher van Bath, *Agrarian Hist. W. Eur. A.D. 500–1850*, 1963, pp. 258–9; 'Studien betreffende de agrarische Geschiedenis van de Veluwe in de Middeleeuwen', *A. A. G. Bijdragen* 1964 xi, 49.

511 Sée, op. cit., 57, 65, 136.

512 Ibid., 37–8, 57, 63, 66, 127, 136; L. Żytkowicz, 'Les Transformations du paysage polonais av. le XIXe s.' in Guarducci, op. cit., 132, 177–8, 183–4, 186–7; Dorošenko, ibid., 208; S. Gissel, 'Vil. and Environment: Ca.: Dan. isls in M.A.', ibid., 304; F. L. Carsten, 'Slavs in N.E. Ger.', *Econ. Hist. Rev.* 1941 xi, 61; Clapham, op. cit., 31, 34–5; Nielsen, op. cit., 181; Schröder-Lembke, 'Wesen', 14sqq.; 'Entstehung u. Verbreitung der Mehrfelderwirtschaft in Nordöstdtschl.', *Zs. f. Agrargesch. u. Agrarsoziologie* 1954 ii, 131; K. S. Bader, *Das Dorf*, 3 vols, Weimar (i, ii), Vienna, Cologne and Graz (iii), 1957–73: i, *Das mittelalterliche Dorf als Friedens- u. Rechtsbereich*, pp. 46–7; Krenzlin and Reusch, op. cit., 14, 111, 113sqq.; Heckscher, op. cit., 26–8, 153; Meitzen, op. cit., i, 69, 70; Wührer, op. cit., 98; *Landw. in Bayern*, 153; Bloch, op. cit., 31, 48, pls IV, V; Lyaschenko, op. cit., 191, 443–5; D. Warriner, *Econ. Peasant Fmg*, 1939, pp. 10, 134, 177; Krzymowski, op. cit., end map; Parain, art. cit., 127; Hanssen, op. cit., i, 139–40, 152 sqq.; Chayanov, op. cit., 139–41; M. Kovalevsky, *Mod. Customs and Anc. Laws Ru.*, 1891, pp. 96, 109, 111; Blum, *Aus.*, 44, 153–4; *Ru.*, 166, 231, 328, 336–7; J. Sion, *Les Paysans de la Normandie orientale*, Paris, 1909, pp. 146, 225, 227; Belényesy, art. cit., 184sqq.; as n. 510.

513 Von Below, op. cit., 42; Blum, *Aus.*, 44, 153–4; *Ru.*, 166; Bloch, op. cit., 33–4; *Cam. Econ. Hist. Eur.*, i (1966), 137.

514 Bader, op. cit., iii: *Rechtsformen u. Schichten der Liegenschaftsnutzung im Mittelalterlichen Dorf*, 94–5; Krenzlin and Reusch, op. cit., 115; Bieleman, op. cit., 580.

515 Ibid., 539, 579; Müller-Wille, 'Schiefergebirge', 559, 564.

516 G. Lefèbrve, *Les Paysans du Nord pendant la Révolution Française*, Bari, 1959, p. 212; O. Festy, *L'Agriculture pendant la Révolution Française*, 2 vols, Paris, 1947–50, i, 14; J. Jacquart, *La Crise rurale en Ile-de-France 1550–1670* Sorbonne Pubs, N.S. Rech. no. 10, Paris, 1974, pp. 292–4; Müller-Wille, *Birkenfeld*, 55–6, 59; Laur, op. cit., 36, 64; Bloch, op. cit., 33–4; H. L. Root, *Peasants and K. in Burgundy: agrarian founds Fr. Absolutism*, Berkeley, Los Angeles and London, 1987, p. 106; Belényesi, art. cit., 184–7; Heckscher, op. cit., 153; Sion, op. cit., 227–8, 345, 347; Musset, art. cit., 143, 150.

517 Ibid., 143; Bloch, op. cit., 33; Heckscher, op. cit., 28, 153; Meitzen, op. cit., i, 69; Sion, op. cit., 146; Chayanov, op. cit., 139–40; L. Delisle, *Etudes sur la condition de la classe agricole et l'état de l'agriculture en Normandie au Moyen Age*, Paris, 1903, pp. 54, 297–8, 300; Parain, art. cit., 127; Steinbach and Becker, op. cit., 42; Żytkowicz, 'Peasant's Fm', 138, 187–8; Zimányi in Guarducci, op. cit., 57, 64; Von Below, op. cit., 42.

518 Ibid.; Krenzlin and Reusch, op. cit., 115; Hard, op. cit., 160–2; Bieleman, op. cit., 539, 574sqq.; Hanssen, op. cit., i, 171, 184sqq.

519 Ibid., 184–5; Irsigler, art. cit., 725–6, 736; B. H. Slicher van Bath in J. H. Bromley and E. H. Kossman, *Brit. and Neth.*, 1960, pp. 133–6, 148; my *Ag. Rev.*, 269–70; Abel, op. cit., 86–7, 217; T. S. Willan, *Tu. Bk Rates*, Manchester, 1962, p. x; B. Dietz, *Pt and Trade E. Eliz. Lond.* Lond. Rec. Soc. 1972, pp. 28, 41–2, 52, 152–4; N. S. B. Gras, *E. Engl. Customs Syst.*, Cambridge, Mass., 1918, p. 109; E. De Laveleye in Probyn, op. cit., 446, 456, 458–9, 465–6; De Vries, op. cit., 71–2, 153–4.

520 Ibid., 141–2; Bieleman, op. cit., 539, 579; Hanssen, op. cit., i, 167, 180–1; Bloch, op. cit., 205–6; Irsigler, art. cit., 720sqq.

521 Ibid., 721, 723, 740–1; Hanssen, op. cit., i, 167, 180; Bader, op. cit., iii, 93, 98; Musset, art. cit., 144.

522 Bloch, op. cit., 29, 41.

523 Müller-Wille, *Birkenfeld*, 87–90.

524 Musset, art. cit., 148–9; Bloch, op. cit., 216, 219; Delisle, op. cit., 325; Jacquart, *Crise*, 292; Festy, op. cit., i, 42; ii, 25sqq.; Sion, op. cit., 225–8, 345, 347, 393, 419.

525 Ibid., 393; L. Aario, 'Die Kulturlandschaft u. bäuerliche Wirtschaft beidenseits des Rheintales bei St Goar', *Acta Geographica* Societas Geographica Fennica, 1945 ix, 24–5; *Landw. in Bayern*, 153; Klag, art. cit., 54; Meyer, art. cit., 132; Bruford, op. cit., 115; Bloch, op. cit., 215, 219, 229; Juillard, *Vie rurale*, 44, 264–6; W. Sombart, *Dtsch. Volkswirtschaft im 19. Jh. u. im Anfang des 20. Jhs*, Berlin, 1927, p. 48; Hömberg, *Siedlungsgesch.*, 94, 99, 100; Nielsen, op. cit., 183–4; Braudel, *Ident.*, ii, 277, 282. 'Ray-grass' is English dialect for 'rye-grass'.

526 Chayanov, op. cit., 143–5; Blum, *Aus.*, 44, 95, 155–6, 168–9; *Ru.*, 337–8; Warriner, *Econ.*, 10, 125–7, 129.

527 Ibid., 10; Juillard, 'L'Assolement', 35, 41; Aario, art. cit., 46–7; *Landw. in Bayern*, 153; Müller-Wille, *Birkenfeld*, 66, 71sqq.; Abel, op. cit., 306sqq.; Mayhew, op. cit., 169–70; Bruford, op. cit., 117; Laur, op. cit., 64; Hömberg, *Siedlungsgesch.*, 94, 99, 100; Nielsen, op. cit., 183; Krzymowski, op. cit., 328sqq.; Blum, *Aus.*, 44, 97, 155–6; Bieleman, op. cit., 539, 574–5; Bloch, op. cit., 215–16; Klag, art. cit., 54; Krenzlin and Reusch, op. cit., 115; Braudel, *Ident.*, ii, 270.

528 Chayanov, op. cit., 142; Warriner, *Econ.*, 125–7, 129, 134.

529 Ibid., 134; *Contrasts*, 308–9, 322; Balassa, art. cit., 111; cf. Thallóczy, op. cit., ii, 14–16.

530 M. L. Ryder and S. K. Stephenson, *Wool Growth*, London and New York, 1968, pp. 144, 167–8, 195–6; Weulersse, op. cit., 152–3, 167–8; Bieleman, op. cit., 481, 487–90; Baden-Powell, *Orig.*, 9; Keating, op. cit., 45; Slater, *Econ. Studs*, 36, 61–2, 65, 78, 88, 99, 106, 113, 131, 172, 187; Sée, op. cit., 79; Meyer, art. cit., 134; De Vries, op. cit., 150; Fei, op. cit., 121; J. Blum, 'Int. Str. and Polity Eur. Vil. Commun. fr. 15th to 19th cent.', *Jnl Mod. Hist.* 1971 xliii, 542; Root, op. cit., 69, 70; B. H. Slicher van Bath, *Mensch en Land in de Middeleeuwen* Academisch Proefschrift, 2 vols, Assen, [1943], ii, 6–8; Bloch, op. cit., 47, 133, 240; Müller-Wille, *Birkenfeld*, 18.

531 Ibid., 12–14, 23, 55–6, 87–90; Donnithorne, op. cit., 34; Fei, op. cit., 37, 39, 237; Weulersse, op. cit., 152–3.

532 Bloch, op. cit., 49.

533 S.C. Powell, op. cit., 94; A.B. MacLear, *Early N.E. Tns*, New York, 1908, pp. 73, 96–8, 110, 130; P. C. Bidwell and J. I. Falconer, *Hist. Ag. in No. U.S. 1620–1860*, Washington, 1925, p. 22; H. L. Osgood, *Am. Cols in 17th cent.*, 3 vols, Gloucester, Mass., 1957, i, 454–6; Thallóczy, op. cit., i, 505–6, 516–17; Bruford, op. cit., 113; Baden-Powell, *Orig.*, 9.

534 Ibid., 13; *Ind. Vil. Commun.*, 278; *Ld Systs*, i, 154–5; MacLear, op. cit., 48, 55, 73, 91–2, 96–8, 104, 106–7, 110–11, 114–16, 129sqq.; Bloch, op. cit., 167sqq.; Lewinski, op. cit., 29–31; Blum, 'Int. Str.', 544–7, 549, 552; Bader, op. cit., pass; Thallóczy, op. cit., i, 505–6, 516–17, 533; S.C. Powell, op. cit., 94, 101–2; Slicher van Bath, *Mensch en Land*, ii, 12, 13.

535 Ibid., 2, 3, 5–8; 'Studien', 64–5; *Agrarian Hist.*, 61; Meyer, art. cit., 132; S.C. Powell, op. cit., 95; Sion, op. cit., 140; Bidwell and Falconer, op. cit., 55; Bieleman, op. cit., 487–9, 577–9, 582sqq., 609–12; Weulersse, op. cit., 100, 107; Bruford, op. cit., 114; G. Franz, *Gesch. des dtsch. Bauernstandes vom frühen Mittelalter bis z. 19. Jh.*, Stuttgart, 1970, pp. 51sqq., 63sqq.; Nielsen, op. cit., 181; Krzymowski, op. cit., 193; Kovalevsky, *Mod. and Anc.*, 111; *Landw. in Bayern*, 644sqq.; Heckscher, op. cit., 27; Warriner, *Ld Ref.*, 43–4; Lyaschenko, op. cit., 444; J. Blum, 'Eur. Vil. as Commun.: origs and functions', *Ag. Hist.* 1971 xlv, 161; 'Int. Str.', 542; *Ru.*, 328, 525; Aario, art. cit., 46–7; Bloch, op. cit., 41–4; Żytkowicz in Guarducci, op. cit., 132, 178; Steinbach and Becker, op. cit., 73sqq., 96, 149; Root, op. cit., 69, 70; Sée, op. cit., 63, 80, 136, 151; Müller-Wille, *Birkenfeld*, 18, 55; Day, loc. cit., 700; cf. Osgood, op. cit., i, 434, 461sqq.; C.P. Nettels, *Roots Am. Civ.*, New York, 1938, p. 235; MacLear, op. cit., 81sqq.

536 Baden-Powell, *Orig.*, 120–2; Osgood, op. cit., i, 433.

537 Warriner, *Econ.*, 118.

538 *Landw. in Bayern*, 653.

539 Steinbach and Becker, op. cit., 34–5, 101, 164; Slicher van Bath, *Agrarian Hist.*, 61; Coopland, op. cit., 123; Irsigler, art. cit., 536.

540 Bloch, op. cit., 43, 237, 240; Juillard, 'L'Assolement', 34.

541 Steinbach and Becker, op. cit., 34–5, 42, 73, 96, 101, 149; Weulersse, op. cit., 100; Heckscher, op. cit., 25–7.

542 As nn. 98–101, 200, 420–2, 431.

543 M. Venard, *Bourgeois et paysans*, Paris, 1957, pp. 119–22; Laur, op. cit., 36; Heckscher, op. cit., 26, 156; Blum, *Aus.*, 149–50; Sion, op. cit., 418; Abel, op. cit., 301–2; Coopland, op. cit., 119; Mayhew, op. cit., 186–7, 189; E. Juillard, A. Meynier, X. de Planhol and G. Sautter, *Structures agraires et paysages ruraux* Ann. de l'Est Mém. 17, p. 78; G. Baer, *Hist. Landownership in Mod. Egy.*, 1962, pp. 79, 81, 83; Warriner, *Econ.*, 132, 160–1; *Contrasts*, 307–8; Krenzlin and Reusch, op. cit., 14, 111, end pkt maps; Keating, op. cit., 39, 40; *Rep. R. Cmmn Ag. Ind.*, 135; Sée, op. cit., 63; Baden-Powell, *Ind. Vil. Commun.*, 314, 414, 437; *Ld Systs*, ii, 136–7; *Orig.*, 76; Meyer, art. cit., 128; J. Jacquart, 'Le Rôle de la grande exploitation dans la formation des paysages des plaines limoneuses de la France du Nord (c.1450 – c.1800)' in Guarducci, op. cit., 46; Żytkowicz, ibid., 178; Ilešič, *Flurformen*, 47; J. Harrison, *Econ. Hist. Mod. Sp.*, Manchester, 1978, p. 5; Weulersse, op. cit., 101, 105–6, 191.

544 Ibid., 191; Fei, op. cit., 33.

545 Baden-Powell, *Ind. Vil. Commun.*, 19; *Ld Systs*, i, 112; Ilešič, *Flurformen*, 47.

546 Slater, *Econ. Studs*, 79, 80; Clapham, op. cit., 49, 50, 202–4; Warriner, *Econ.*, 161; Jacquart, 'Le Rôle', 47; Mayhew, op. cit., 187; Abel, op. cit., 80, 302–3; Juillard, 'L'Assolement', f.p. 35; Heckscher, op. cit., 156; G.T. Robinson, *Rural Ru. under O. Regime*, 1932, pp. 219sqq.; Weulersse, op. cit., 101; Keating, op. cit., 41; *Rep. R. Cmmn Ag. Ind.*, 138; Sée, op. cit., 66; Laur, op. cit., 36; Müller-Wille, 'Langstreifenflur', 12; Parain, art. cit., 134; *Landw. in Bayern*, 653–4, 698sqq.

547 Ibid., 654, 698sqq.; Harrison, op. cit., 159–60; Dipl. and Consr Reps, *Ger.: Rep. on Ag. in Ger.*, 1898, p. 9; Clapham, op. cit., 49, 50, 202–3; Mayhew, op. cit., 187sqq.; T.C. Smith, op. cit., 19; L.E. Textor, *Ld Ref. in Cz.*, 1923, pp. 110sqq.; Juillard et al., op. cit., 30; Weulersse, op. cit., 191; C. Christians, 'Quelques aspects de la structure agraire et de l'aménagement rural en Ardennes Belges' in *Géog. et Hist. Agraires* Ann. de l'Est Mém. 21, pp. 83–5; H.C. Wallich, *Mainsprings Ger. Revival*, New Haven, 1955, p. 126; *Outlines Ag. Jap.*, 22, 116; H.G. Wanklyn, *E. Marchlands Eur.*, 1941, p. 226; Schröder, art. cit., 328–9; Heckscher, op. cit., 156.

548 Ibid., 161.

549 Ibid., 192–3.

550 Warriner, *Econ.*, 16; *Contrasts*, 83–4; Clapham, op. cit., 49, 201, 204; Abel, op. cit., 301; Bloch, op. cit., 222, 232.

551 Ibid., 46; Baden-Powell, *Ind. Vil. Commun.*, 49, 50.

552 Lyaschenko, op. cit., 100, 180; Weulersse, op. cit., 145–6.

553 W.G. Simkhowitsch, *Die Feldgemeinschaft in Russland*, Jena, 1898, pp. 3, 28sqq., 39, 85; A. Hömberg, *Die Entstehung der westdtsch. Flurformen: Blockgemengflur, Streifenflur, Gewannflur*, Berlin, 1935, pp. 14, 15; J. Hövermann, *Die Entwicklung der Siedlungsformen in der Marschen des Elbe-Wesel Winkels* Forschungen z. dtsch. Landeskunde, Bd 56, Remagen, 1951, pp. 17, 69, maps 1–9, 11–14, 19, 22–3; Steinbach, art. cit., 19–21, 24sqq.; Mayhew, op. cit., 24, 67, 73; Schröder, art. cit., 310–13, 315, 320; Bloch, op. cit., pl. I; Krenzlin, 'Blockflur', 353sqq., 364–5; 'Z. Genese', 193–4, 200–1, 203; J.W. Thompson, *Feud. Ger.*, Chicago, 1928, p. 512; Ilešič, *Flurformen*, bet. 16–17, 24, 26–7, 61sqq., 71–2, 75, 79, 123–4; 'Gewannfluren', 131; Müller-Wille, 'Langstreifenflur', 15, 16, 25, 27–30.

554 Ibid., 28; Ilešič, *Flurformen*, 24, 26–7, 52–4, 73, 75; Thompson, op. cit., 475, 477, 485, 503, 509–10, 573; Krenzlin, 'Probleme', 547sqq., 556.

555 Lewinski, op. cit., 23, 49.

556 T.C. Smith, op. cit., 4, 7, 106; Donnithorne, op. cit., 53; Baden-Powell, *Orig.*, 22–4; Phiillpotts, op. cit., 125sqq., 271; Bloch, op. cit., 150sqq.; Weulersse, op. cit., 216–17, 219; Blum, *Ru.*, 24–5, 515; 'Int. Str.', 563–4; Warriner, *Contrasts* 2, 257–60, 320, 369; Kovalevsky, *Mod. and Anc.*, 46–7, 67, 75–6, 78–80; G. Vernadsky, *Kievan Ru.* (Vernadsky and M. Karpovich, *Hist. Ru.*, vol. ii), New Haven, 1948, pp. 133, 135; Baer, op. cit., 79; P. Laslett, *Hsehld and Fam. in Past Time*, Cambridge, 1972, pp. 14sqq.; Lyaschenko, op. cit., 63, 67; Simkhowitsch, op. cit., 8sqq., 23, 25–6, 39, 40, 349sqq., 362sqq.; Robinson, op. cit., 118.

557 Ibid.; T.C. Smith, op. cit., 106, 128–9, 145–6; Donnithorne, op. cit., 53–4, 56–9; Warriner, *Contrasts*, 2, 260, 342; Kovalevsky, *Mod. and Anc.*, 65–7, 77–8; Blum, *Ru.*, 26; Baer, op. cit., 79; Vernadsky, op. cit., 109; Baden-Powell, *Orig.*, 24, 104–5; *Ind. Vil. Commun.*, 257; Ilešič, *Flurformen*, 72.

558 Lewinski, op. cit., 29, 30, 50; T. C. Smith, op. cit., 184–7; Kovalevsky, *Mod. and Anc.*, 77–8, 92–3; Vernadsky, op. cit., 133; Lyaschenko, op. cit., 69–71, 89; Simkhowitsch, op. cit., 22, 27, 29–31, 61sqq., 67–8, 70–1, 90sqq., 103–4.

559 Gissel, loc. cit., 304; Thompson, op. cit., 494, 508, 512; Bloch, op. cit., 150sqq.; Parain, art. cit., 127–8, 130; Slicher van Bath, *Mensch en Land*, i, 203, 213sqq., 293sqq.; Hömberg, op. cit., 30–1.

560 Ibid., 31; Slicher van Bath, *Agrarian Hist.*, 77–9; Ilešič, *Flurformen*, 72; Lewinski, op. cit., 16; cf. Krenzlin and Reusch. op. cit., 102–5, 111.

561 Ilešič, 'Gewannfluren', 131–4.

562 Blum, *Ru.*, 7, 80, 89, 93, 113–14, 219–21, 226, 245, 251, 254–5, 263–4, 270–1, 297–9, 308–10, 318–20, 405, 413, 415–16, 420, 422, 424–7, 434, 455–7, 468–9, 593, 596–8, 618–19; B. H. Sumner, *Surv. Ru. Hist.*, 1944, pp. 142, 145sqq., 152–3, 160; I. L. Evans, *Agrarian Rev. in Roumania*, Cambridge, 1924, pp. 19–21, 38, 45, 156; F. Lot, *End Anc. Wld and Begs M. A.*, 1931, pp. 107, 109–12; Kovalevsky, *Mod. and Anc.*, 106–8, 211sqq.; Ilešič, 'Gewannfluren', 136; T. C. Smith, op. cit., 17, 19, 23, 131, 134–5.

563 Weulersse, op. cit., 123.

564 Baden-Powell, *Orig.*, 95; *Ind. Vil. Commun.*, 324–5; *Ld Systs*, ii, 679; Bloch, op. cit., 45.

565 Clapham, op. cit., 34; J. Mavor, op. cit., i, 46; Thompson, op. cit., 485, 501–2, 508, 510–12; Krenzlin, 'Blockflur', 357–9, 365; Żytkowicz in Guarducci, op. cit., 137–8, 183–4; M. K. Zaleska, 'L'Ancien Morcellement des champs av. la séparation au XIXe s. dans la Poméranie de Gdansk' in *Géog. et Hist. Agraires* Ann. de l'Est Mém. 21, pp. 345sqq.

566 S. C. Powell, op. cit., 79, 81, 95, 109; Nettles, op. cit., 235; Osgood, op. cit., i, 433, 439sqq., 449sqq., 458–60; MacLear, op. cit., 34, 81, 84–7, 89; E. Scofield, 'Orig. Set. Patts in Rural N.E.', *Geog. Rev.* 1938 [xxviii], 652sqq.; G. T. Trewartha, 'Types Rural Set. in Col. Am.', ibid. 1946 xxxvi, 568, 570sqq., 587, 591; H. B. Adams, 'Ger. Origs N.E. Tns' in Adams, *Loc. Insts* ed. H. B. Adams, Johns Hopkins Univ. Studs vol. i, Baltimore 1883 (1882–3), 32sqq.; Bidwell and Falconer, op. cit., 21–2, 51, 55, 58.

567 Ibid., 58; MacLear, op. cit., 81–2, 84sqq.; Scofield, art. cit., 656–7, 660–1; Trewartha, art. cit., 570–1; Blum, *Ru.*, 527; Weulersse, op. cit., 123; T. C. Smith, op. cit., 19; Warriner, *Ld Ref.*, 43–4; Lot, op. cit., 109; Parain, art. cit., 128, 130; Baden-Powell, *Ind. Vil. Commun.*, 250, 254–5, 280–1, 283, 291–2, 334, 414; *Ld Systs*, ii, 136–7, 140–3; *Orig.*, 93; Osgood, op. cit., i, 458–60; J. Mavor, op. cit., i, 209, 211, 215; ii, 259.

568 Krenzlin and Reusch, op. cit., 94, 110; Ilešič, *Flurformen*, 123–4; 'Gewannfluren', 136; Krenzlin, 'Probleme', 567.

569 Baden-Powell, *Orig.*, 104–5, 143sqq.; *Ind. Vil. Commun.*, 257.

570 Weulersse, op. cit., 42, 99–102, 104–6, 189–92, 216–17; Warriner, *Ld Ref.*, 98; Ilešič, 'Gewannfluren', 130.

571 Ibid.

572 Lewinski, op. cit., 13sqq.; Robinson, op. cit., 11, 12, 34–5, 39, 113, 120–2, 213–14; Kovalevsky, *Mod. and Anc.*, 66–7, 78–80, 93–6, 109–10, 113–15; J. Mavor, op. cit., i, 208–9, 211, 215; ii, 259, 274–5, 341; Sumner, op. cit., 132–3; Simkhowitsch, op. cit., 29, 30, 77sqq., 103–4, 122sqq., 136sqq.; Blum, *Ru.*, 508sqq., 516, 519–22, 527–8.

573 Ibid., 527; MacLear, op. cit., 84; Bidwell and Falconer, op. cit., 50, 52, 57–8; Baden-Powell, *Ld Systs*, ii, 136–7, 143, 679; *Ind. Vil. Commun.*, 414; Weulersse, op. cit., 99, 105, 123; Jutikkala, art. cit., 122–3, 126, 131, 134–6; Thompson, op. cit., 484–5, 512; Krenzlin and Reusch, op. cit., 79; Schröder, art. cit., 314, 317–18, 323; Ilešič, 'Gewannfluren', 133, 136–7.

574 Mortensen, op. cit., i, 97; ii, 43; Ilešič, *Flurformen*, 54, 56.

575 Slicher van Bath, *Mensch en Land*, ii, 18, 19.

576 Von Below, op. cit., 41, 69, 70; Lamprecht, op. cit., I.i, 545; Bloch, op. cit., 33–4; Ilešič, *Flurformen*, 54; cf. L. White, *Med. Tech. and Soc. Change*, Oxford, 1962, pp. 69, 70.

577 H. Wartmann, *Urkundenbuch der Abtei Sanct Gallen*, 2 vols, Zurich, 1863–6, i, 41 (no. 39); O. Dobenecker, *Regesta Diplomatica necnon Epistolaria Historiae Thuringiae*, 2 vols, Jena, 1896 i, 15, 19 (nos 48, 66); cf. Krenzlin, 'Blockflur', 357; Seebohm, op. cit., 318sqq.; pace ibid. 321; L. White, op. cit., 69.

578 F. Lütge, *Die Agrarverfassung des frühen Mittelalters in mitteldtsch. Raum, vornehmlich in der Karolingerzeit*, Jena, 1937, p. 293; cf. Slicher van Bath, *Mensch en Land*, i, 203.

579 R. Latouche, *Bth W. Econ.*, 1961, pp. 191, 193–4; Irminon, *Polyptyque de l'Abbé Irminon, ou dénombrement des manses, des serfs et des revenus de l'abbaye de Saint-Germain-des-Prés* ed. B. E. C. Guérard, 2 vols, Paris, 1844, ii, 78; cf. Bloch, op. cit., 34.

580 Irminon, op. cit., ii, 4, 5, 77, 95–7, 105, 107sqq., 119–20, 131, 138, 145, 147–8, 191sqq.

581 Ibid., 126–8.

582 Ibid., 52, 56, 63, 224, 259.

583 Ibid., 2, 6, 29, 33, 38, 41, 44, 47, 62, 66–7, 70–1, 73–4, 77, 109, 151sqq., 165sqq., 174, 179–81, 208sqq., 214sqq., 236, 239, 246sqq., 272sqq.

584 Ibid., 182sqq., 228sqq., 241sqq., 269.

585 Ibid., 113, 132, 224, 254, 259, 267, 269.

586 Ibid., 52, 119.

587 Ibid., 63sqq., 183, 221–2, 224.

588 Ibid., i, 170; ii, 24sqq.

589 Ibid., 132sqq.; for spelt, ibid. 131sqq., esp. 149; cf. ibid., i, 709; ii, 76–8, 97–8, 107, 110–11, 208sqq.

590 Ibid., i, 677–80 (quot. p. 679); ii, 38, 41, 44, 47, 151, 164–5, 182, 274, 452, 456; for porkers, carnaticum and shingles, pp. 5, 151, 182. For carnaticum, per indice. Pace Latouche, op. cit., 194.

591 Bloch, op. cit., 28–9, 31–4; Lewis, op. cit., 198.

592 Festy, op. cit., i, 14; ii, 9, 11; cf. C. Higounet, *Structure et exploitation d'un terroir Cistercien de la plaine de France XIIe–XVe s.*, Paris, 1965, pp. 42–3.

593 G. Duby, 'La Révolution agricole médiévale', *Revue de Géog. de Lyon* 1954 xxix, 361–2.

594 G. Duby, *L'Econ. rurale et la vie des campagnes dans l'occ. méd.: Fr., Angl., Emp. IXe–XVe s.*, 2 vols, Paris, 1962, i, 81.
595 Irminon, op. cit., i, 925; cf. Duby, *L'Econ.*, i, 81.
596 Ibid., i, 178.
597 J.R. Strayer, 'Econ. Conditions in Co. Beaumont-le-Roger 1261–1313', *Speculum* 1951, xxvi, 277sqq., esp. 280.
598 J.R. Strayer, *R. Domain in Bailliage Rouen*, 1976, p. 30.
599 Irminon, op. cit., i, 925.
600 Ibid.
601 Ibid., 926.
602 Irsigler, art. cit., 523.
603 G. Duby, *Early Growth Eur. Econ.: warriors and peasants fr. 7th to 12th cent.*, Ithaca, N.Y., 1974, p. 191.
604 C. Higounet, 'L'Assolement triennal dans la plaine de France au XIIIe s.', Acad. des Inscriptions et Belles-Lettres, *Comptes rendus des séances de l'année 1956*, Paris, 1956, pp. 507sqq.; *Structure*, 42sqq., 69, figs 16, 17.
605 Ibid., 42–3.
606 Ibid., 42; J.J. Hoebaux, 'Encore et toujours à propos de assolement triennal au Moyen Age', *Revue Belge de la Philologie et d'Hist.* 1966 xliv, 1174–6.
607 Sion, op. cit., 146; Delisle, op. cit., 298, 300.
608 Musset, art. cit., 143, 150.
609 Coopland, op. cit., 123.
610 Ibid., 121–3; Sion, op. cit., 146, 225; Jacquart, *La Crise*, 292; Higounet, 'L'Assolement', 509; Bloch, op. cit., 206–9; G. Fourquin, *Les Campagnes de la région Parisienne à la fin du Moyen Age*, Paris, 1964, pp. 76sqq.
611 Ibid., 191–2, 225sqq., 290sqq.
612 Higounet, 'L'Assolement', 509.
613 Irsigler, art. cit., 720sqq.
614 Hanssen, op. cit., i, 166.
615 Müller-Wille, 'Langstreifenflur', 15, 16; 'Schiefergebirge', 558, 560, 564.
616 Ibid., 562, 564; *Birkenfeld*, 55–6, 96sqq., 102–3; Krenzlin and Reusch, op. cit., 56, 97–8, 101, 105–6, 111, 113sqq.; Krenzlin, 'Probleme', 547sqq., 556, 567; Steinbach and Becker, op. cit., 42, 149; *Landw. in Bayern*, 653; Hömberg, *Siedlungsgesch.*, 93–4, maps 5, 5a; Hanssen, op. cit., i, 166–7; Braudel, *Ident.*, ii, 355; Mayhew, op. cit., 85, 91sqq., 118sqq., 169; Abel, op. cit., 110sqq., 150sqq., 201sqq.
617 Heckscher, op. cit., 26, 153; Johnsen, op. cit., 408; Gissel, art. cit., 304; Wührer, op. cit., 98; Jutikkala, art. cit., 119sqq.; Nielsen, op. cit., 125, 146, 181.
618 D. Warriner, 'Some Controversial Issues in Hist. Agrarian Eur.', *Slav. and E. Eur. Rev.* 1953 xxxii, 175; Żytkowicz in Guarducci, op. cit., 138, 177–8, 186–8.
619 Ibid., 138, 183–4; Dorošenko, ibid., 208; Zaleska, art. cit., 345sqq.; K. von Loewe, 'Comm. and Ag. in Lith. 1400–1600', *Econ. Hist. Rev.* 2nd ser. 1973 xxvi, 28–9; R.A. French, '3-Fld Syst. 16th cent. Lith.', *Ag. Hist. Rev.* 1970 xviii, 107–8, 118–19; Braudel, *Ident.*, ii, [356].

620 Ibid.; Blum, *Ru.*, 22–3, 26, 119, 166, 231, 336–7; Chayanov, op. cit., 139–41; Lyaschenko, op. cit., 179–80, 191, 444–5; Robinson, op. cit., 11, 34; Vernadsky, op. cit., ii, 107–9; J. Mavor, op. cit., i, 211; ii, 274; Lewinski, op. cit., 29.

621 Müller-Wille, 'Schiefergebirge', 565.

622 Mayhew, op. cit., 27; Krenzlin and Reusch, op. cit., 111, 113–15, 117.

623 Ibid., 24–6, 30, 32, 47, 49, 78–9, 83, 91, 97–8, 101–4, 106, 111, 113–16; Müller-Wille, 'Langstreifenflur', 21, 23–4, 43–4; *Birkenfeld*, 13, 14, 96sqq., 124; Ilešič, *Flurformen*, 54–6; Krenzlin, 'Blockflur', 355–6; 'Z. Genese', 191, 193–5, 198, 201, 203; Abel, op. cit., 75sqq., 150sqq., 201sqq.; Mayhew, op. cit., 26; Hömberg, *Entstehung*, 23, 50–1, 57; Steinbach and Becker, op. cit., 42.

624 Blum, *Ru.*, 527.

625 Ilešič, 'Gewannfluren', 131sqq.; *Flurformen*, 58–9, 61sqq., 71–2, 74–6.

626 My *Text. Manufs in E. Mod. Engld*, Manchester, 1985, pp. 2, 32; A. J. Bourde, *Infl. Engld on Fr. Agronomes 1750–89*, Cambridge, 1953, pp. 131, 135sqq., 142, 209–12; J. Jacquart, 'Fr. Ag. in 17th cent.' in P. Earle, *Essays in Eur. Econ. Hist. 1500–1800*, Oxford, 1974, pp. 170–1, 173; J. Smith, *Chronicon Rusticum-Commerciale or Mems Wl etc.*, 2 vols, 1747, ii, 135, 419.

627 Phillpotts, op. cit., 49, 65–7, 76sqq., 202–4, 245–6; Slicher van Bath, *Mensch en Land*, i, 213–16; ii, 3, 5, 7, 8, 12.

628 Hömberg, *Entstehung*, 29–31; Thompson, op. cit., 494; Bloch, op. cit., 160–3; Braudel, *Ident.*, i, 104–5.

629 Ibid.; Warriner, *Contrasts*, 257–60, 297, 320, 342, 344, 369; Ilešič, 'Problèmes', 290–1; *Flurformen*, 72; Blum, *Ru.*, 26; 'Eur. Vil.', 169–70; 'Int. Str.', 565; A. Blanc, 'Communautés rurales et strs agraires dans les pays Sud-slaves' in *Géog. et Hist. Agraires* Ann. de l'Est Mém. 21, pp. 57sqq., 65; Simkhowitsch, op. cit., 8sqq., 25–7, 31, 349sqq., 362sqq.; Kovalevsky, *Mod. and Anc.*, 66–7, 78–80.

630 Ibid., 106–8, 211sqq.; R. E. F. Smith, *Enserfment Ru. Peasantry*, Cambridge, 1968, p. 27; Blum, *Ru.*, 114, 219–20, 226, 420, 593, 596–8, 618–19; Evans, op. cit., 19–21, 38, 45, 156; Robinson, op. cit., 118; Simkhowitsch, op. cit., 19sqq., 61sqq., 67sqq.; French, art. cit., 111–12, 118, 123.

631 Krenzlin and Reusch, op. cit., 115; Bader, op. cit., i, 46; Krzymowski, op. cit., 191.

632 Hömberg, *Entstehung*, 38; Krenzlin, 'Z. Genese', 193.

633 Tacitus, *Germania*, cap. 26: *Agri pro numero cultorum ab universis in vices occupantur, quos mox inter se secundum dignationem partiuntur; facilitatem partiendi camporum spatia praestant. Arva per annos mutant, et superest ager. Nec enim cum ubertate et amplitudine soli labore contendunt, ut pomaria conserant ut prata separent ut hortos rigent.* Cf. Abel, op. cit., 17, 18; Hanssen, op. cit., i, 127–9; Latouche, op. cit., 34–5.

634 Haudricourt and Delamarre, op. cit., 355, 358–9, 363; Steensberg, art. cit., 91, 96; Sandred, art. cit., 259; Müller-Wille, 'Langstreifenflur', 29, 30.

635 Ibid., 25; Haudricourt and Delamarre, op. cit., 349, 355; Hövermann, op. cit., 15; Thompson, op. cit., 471–2, 485, 490–1, 494–6, 501sqq., 508–10,

521, 532–3, (f.p. 545), 545–7, 552sqq., 560sqq., 579, 581sqq., 589sqq., 607, 612sqq., 657–8; Carsten, art. cit., 61, 64; Krenzlin, 'Probleme', 547sqq., 556–8, 561; 'Blockflur', 357–9, 365; Ilešič, *Flurformen*, 124; Warriner, 'Issues', 174–5; Żytkowicz in Guarducci, op. cit., 137–8, 176, 183–4, 186; Zaleska, art. cit., 345sqq.; Mortensen, op. cit., i, 22sqq., 63sqq., 97; ii, 43; H. Pirenne, *Econ. and Soc. Hist. Med. Eur.*, 1936, pp. 77–8; Mayhew, op. cit., 50sqq.; F. Lütge, *Gesch. der dtsch. Agrarverfassung vom frühen Mittelalter bis z. 19. Jh.*, Stuttgart, 1967, pp. 111sqq.

636 Haudricourt and Delamarre, op. cit., 352, 358–9, 363.

637 As n. 626.

638 My *Ag. Rev.*, 43–5, 51–2, 55, 58–9, 62–4, 66–9, 71–2, 74sqq., 81, 88, 96, 114, 117, 119, 121, 142–3, 149, 174, 206, 311–13; *Text. Manufs*, 2, 13; Lewis, op. cit., 206, 298, 302–3, 333, 420–2.

639 Ibid., 206, 223–4, 298, 302–5, 331sqq., 424sqq., 474; P.D.A. Harvey, 'Engl. Infln 1180–1220' in Hilton, *Peasants, Knts and Heretics*, 79–82; Smith, *Mems Wl*, i, 21–2; R. De Roover, *Gresham on For. Ex.*, Cambridge, Mass., 1949, pp. 31sqq.; P. Chorley, 'Engl. Cl. Exps during 13th and 14th cents.: contl evid.', *Hist. Res.* 1988 lxi, 1sqq.; F. Barlow, *Ed. Conf.*, 1970, pp. 181–4; Stenton, *A.-S. Engld*, 221–2, 412–13; Poole, *Domesday*, 84, 90–1; M. Powicke, *13th Cent. 1216–1307*, Oxford, 1962, p. 622; M. McKisack, *14th Cent. 1307–99*, Oxford, 1959, p. 350; E.F. Jacob, *15th Cent. 1399–1485*, Oxford, 1961, p. 346; M. Prestwich, *Three Edwards: war and state in Engld 1272–1377*, 1980, pp. 236–8.

640 Ibid., 27sqq., 33, 72, 106–7, 172, 215–17, 223–4, 235–7, 243, 267–8; McKisack, loc. cit.; Lewis, op. cit., 424–5; Smith, *Mems Wl*, i, 16sqq.

Select bibliography

Place of publication is London unless otherwise stated.

Anonymous or various

Agricultural Bureau, Department of Agriculture and Commerce, *Outlines of Agriculture in Japan*, Tokyo, 1910.

Annales de l'Est Mémoire 17 *Structures Agraires et Paysages Ruraux*, Nancy, 1957.

—— Mémoire 21 *Géographie et Histoire Agraires*, Nancy, 1959.

Descriptive Catalogue of the Manorial Rolls belonging to Sir H. F. Burke Manorial Society Publication xii, 2 pts, 1923.

Die Landwirthschaft in Bayern, Munich, 1890.

Diplomatic and Consular Reports, *Germany: Report on Agriculture in Germany*, 1898.

House of Lords, *Manuscripts of the House of Lords*, new series, several vols, 1900– in progress.

Manorial Society, see *Descriptive Catalogue.*

Ministry of Agriculture and Fisheries, *Dutch Agriculture*, The Hague, n.d.

'Proceedings in the Court of Star Chamber temp. Henry VIII and Edward VI', *Collections for the History of Staffordshire*, William Salt Archaeological Society, 1910, 1912.

Public Record Commissioners, *The Ancient Laws and Institutions of England*, 1840.

Record Commission, *Calendars of the Proceedings in Chancery in the reign of Queen Elizabeth*, 3 vols, 1827–32.

—— *Placitorum in Domo Capitulare Westmonasteriensi Asservatorum Abbrevatio temporibus Ric. I, Johann., Henr. III, Ed. I, Ed. II*, 1811.

—— *Rotuli Hundredorum temp. Hen. III et Edw. I in Turr. Lond. et in Curia Receptae Scaccarii West. asservati*, 2 vols, 1812–18.

Royal Commission on Agriculture in India, *Report of the Royal Commission on Agriculture in India*, (Cmd. 3132), 1928.

Shotley Parish Records, Suffolk Green Books no. xvi, Bury St Edmunds, 1912.

Victoria History of the Counties of England.

Wills and Inventories illustrative of the history, manners, language, statistics etc. of the Northern Counties of England from the Eleventh Century downwards, pt i, Surtees Society, 1835.

Aario, L. 'Die Kulturlandschaft und bäuerliche Wirtschaft beiderseits des Rheintales bei St Goar', *Acta Geographica*, ix, Societas Geographica Fennia, Helsinki, 1945.

Abel, W. *Geschichte der deutschen Landwirtschaft vom frühen Mittelalter bis zum 19. Jahrhundert*, Stuttgart, 1967.
Adams, H. B. 'The Germanic Origin of New England Towns', in *Johns Hopkins University Studies*, ed. H. B. Adams, vol. i, *Local Institutions*, Baltimore, 1883 (1882–3).
Allison, K. J. 'Flock Management in the Sixteenth and Seventeenth Centuries', *Economic History Review*, 2nd series, xi, 1958.
—— 'The Sheep-Corn Husbandry of Norfolk in the Sixteenth and Seventeenth Centuries', *Agricultural History Review*, v, 1957.
—— 'The Wool Supply and the Worsted Cloth Industry in Norfolk in the Sixteenth and Seventeenth Centuries', typescript thesis, Ph.D., Leeds University, 1955.
Appleby, A. B. *Famine in Tudor and Stuart England*, Liverpool, 1978.
Ashby, M. K. *The Changing English Village: a history of Bledington, Gloucestershire, in its setting, 1066–1914*, Kineton, 1974.
Aspley Heath School Historical Society, *A History of Our District*, Aspley Guise, 1931.
Aston, T. S. 'The Origins of the Manor in England', *Transactions of the Royal Historical Society*, 5th series, viii, 1958.
Atwell, G. *The Faithfull Surveyour*, Cambridge, 1662.
Aubrey, J. *An Essay towards the Description of the North Division of Wiltshire*, sine loco, 1838.
—— 'An Introduction to the Survey and Natural History of the North Division of the County of Wiltshire' in *Miscellanies on Several Curious Subjects* (E. Curll), 1714.
—— *The Natural History of Wiltshire*, ed. by J. Britton, 1847.
—— *Wiltshire: the topographical collections of John Aubrey, F.R.S., A.D. 1659–70*, ed. J. E. Jackson, Devizes, 1862.
Ault, W. O. (ed.) *Court Rolls of the Abbey of Ramsey and of the Honor of Clare*, New Haven, 1928.
—— *Open-Field Farming in Medieval England: a study in village by-laws*, 1972.
—— *Open-Field Husbandry and the Village Community: a study of agrarian by-laws in medieval England*, Transactions of the American Philosophical Society, new series, lv, pt vii, 1965.
—— *Private Jurisdiction in England* Yale Historical Publications, Miscellany x, New Haven, 1923.
—— 'Some Early English Village By-laws', *English Historical Review*, xlv, 1930.
—— 'The Manor Court and the Parish Church in Fifteenth-Century England: a study in village by-laws', *Speculum*, xlii, 1967.
—— 'Village By-laws by Consent', *Speculum*, xxix, 1954.
Austin, W. *The History of Luton*, 2 vols, Newport, I. of Wight, 1928.

Baden-Powell, B. H. *The Indian Village Community*, 1896.
—— *The Land Systems of British India*, 3 vols, Oxford, 1892.
—— *The Origin and Growth of Village Communities in India*, 1899.
Bader, K. S. *Das Dorf*, 3 vols: i, *Das mittelalterliche Dorf als Friedens- und Rechtsbereich*, Weimar, 1957; ii, *Dorfgenossenschaft und Dorfgemeinde*, Weimar, 1962;

iii, *Rechtsformen und Schichten der Liegenschaftsnutzung im mittelalterlichen Dorf*, Vienna, Cologne and Graz, 1973.
Baer, G. *History of Landownership in Modern Egypt 1800–1950*, 1962.
Baigent, F.J. and Millard, J.E. *A History of the Ancient Town and Manor of Basingstoke*, Basingstoke, 1889.
Bailey, J. *A General View of the Agriculture of the County of Durham*, 1810.
Bailey, M. *A Marginal Economy? East Anglian Breckland in the later Middle Ages*, Cambridge, 1989.
—— 'Sand into Gold: evolution of the foldcourse in W. Suffolk, 1200–1600', *Agricultural History Review*, xxxviii, 1990.
Baker, A. R. H. 'Field Systems in the Vale of Holmesdale', *Agricultural History Review*, xiv, 1966.
—— 'The Field Systems of Southeast England' in Baker and Butlin, q.v.
—— 'Open Fields and Partible Inheritance on a Kentish Manor', *Economic History Review*, 2nd series, xvii, 1964.
—— 'Some Fields and Farms in Medieval Kent', *Archaeologia Cantiana*, lxxx, 1965.
—— 'The Field Systems of Kent', typescript thesis, Ph.D., London University, 1963.
Baker, A. R. H. and Butlin, R. A. (eds) *Studies of Field Systems in the British Isles*, Cambridge, 1973.
Balassa, I. 'Der Maisbau in Ungarn', *Acta Ethnographica Academiae Scientiarum Hungaricae*, tomus v, fasciculi 1, 2, Budapest, 1956.
Ballard, A. 'Notes on the Open Fields of Oxfordshire', Oxfordshire Archaeological Society's *Report*, 1908 (1909).
—— 'Tackley in the Sixteenth and Seventeenth Centuries', Oxfordshire Archaeological Society's *Report*, 1911 (1912).
Barger, E. 'The Present Position of Studies in English Field Systems', *English Historical Review*, liii, 1938.
Barley, M.W. 'East Yorkshire Manorial Bye-laws', *Yorkshire Archaeological Journal*, xxxv, 1943.
Barlow, F. *Edward the Confessor*, 1970.
Barratt, D. M. (ed.) *Ecclesiastical Terriers of Warwickshire Parishes*, 2 vols, Dugdale Society Publications, xxii, xxvii, 1955–71.
Barrett-Lennard, T. 'Two Hundred Years of Estate Management at Horsford during the Seventeenth and Eighteenth Centuries', *Norfolk Archaeology*, xx, 1921.
Batchelor, T. *A General View of the Agriculture of the County of Bedford*, 1808.
Bateson, E., Hinds, A. B., Hodgson, J. C., Craster, H. H. E., Vickers, K. H., and Dodds, M. W. *A History of Northumberland*, 15 vols, 1893–1940.
Bateson, M. (ed.) *Records of the Borough of Leicester*, vol. iii, Cambridge, 1905.
Batho, G. R. (ed.) *The Household Papers of Henry Percy, Ninth Earl of Northumberland (1564–1632)*, Royal Historical Society, Camden 3rd series, xciii, 1962.
Reale, J. *Herefordshire Orchards, A Pattern for all England*, 1657.
Beckwith, I. 'The Remodelling of a Common Field System', *Agricultural History Review*, xv, 1967.
Bede (Ven.) *Historiae Ecclesiasticae Gentis Anglorum* ed. J. Smith, Cambridge, 1722.

Belényesy, M. 'Angaben über die Verbreitung der Zwei- und Dreifelderwirtschaft im mittelalterlichen Ungarn', *Acta Ethnographica Academiae Scientiarum Hungaricae*, tomus v, fasciculi 1, 2, Budapest, 1956.

Below, G. von, *Geschichte der deutschen Landwirtschaft des Mittelalters in ihren Grundzügen*, Jena, 1937.

Bennett (Canon) 'The Orders of Shrewton', *Wiltshire Archaeological Magazine*, xxiii, 1887.

Bennett, H. S. *Life on the English Manor*, Cambridge, 1938.

Beresford, M. W. 'Lot Acres', *Economic History Review*, xiii, 1941–3.

Bidwell, P. C. and Falconer, J. I. *History of Agriculture in the Northern United States 1620–1860*, Washington, D.C., 1925.

Bieleman, J. *Boeren op het Drentse Zand 1600–1910: een nieuwe visie op het 'oude' landbouw*, being *A. A. G. Bijdragen* xxix, Wageningen, 1987.

Billingsley, J. *A General View of the Agriculture in the County of Somerset*, 1794, 1798.

Billson, C. J. 'The Open Fields of Leicester', *Transactions of the Leicestershire Archaeological Society*, xiv, 1925–6.

Birch, W. de G. (ed.) *Cartularium Saxonicum*, 3 vols, 1885–93.

Bishop, T. A. M. 'Assarting and the Growth of the Open Fields', *Economic History Review*, vi, 1935, reprinted in E. M. Carus-Wilson (ed.) *Essays in Economic History*, [i] q.v.

—— 'The Rotation of Crops at Westerham', *Economic History Review*, ix, 1938–9.

B[lagrave], J. *The Epitome of the Art of Husbandry*, 1669.

Blake, E. O. (ed.) *Liber Eliensis* Royal Historical Society, Camden 3rd series, xcii, 1962.

Blanc, A. 'Communautés rurales et structures agraires dans les pays Sud-slaves', in *Géographie et Histoire Agraires*, Annales de l'Est, q.v.

Bland, A., Brown, P. and Tawney, R. H. (eds) *English Economic History: select documents*, 1925.

Blith, W. *The English Improver Improved*, 1652.

Bloch, M. *French Rural History: an essay on its basic characteristics*, 1966.

Blome, R. *Britannia*, 1673.

Blomefield, F. *An Essay towards a Topographical History of the County of Norfolk*, 5 vols, Lynn, 1775.

Blum, J. *Lord and Peasant in Russia from the Ninth to the Nineteenth Century*, Princeton, N.J., 1961.

—— *Noble Landowners and Agriculture in Austria 1815–1848*, Baltimore, 1947.

—— 'The European Village as Community: origins and functions', *Agricultural History*, xlv, 1971.

—— 'The Internal Structure and Polity of the European Village Community from the Fifteenth to the Nineteenth Century', *Journal of Modern History*, xliii, 1971.

Blundell, J. H. *Toddington: its Annals and People*, Toddington, 1925.

Bodington, E. J. 'The Church Surveys in Wiltshire 1649–50', *Wiltshire Archaeological Magazine*, xl, xli, 1917–19, 1920–2.

Bouch, C. M. L. and Jones, G. P. *A Short Economic and Social History of the Lake Counties, 1500–1830*, Manchester, 1961.

Bourde, A.J. *The Influence of England on the French Agronomes 1750–89*, Cambridge, 1953.
Bowen, H.C. *Ancient Fields: a tentative analysis of vanishing earthworks and landscapes*. British Association for the Advancement of Science, 1961.
Boyd, A.W. *A Country Parish*, 1951.
Boys, J. *A General View of the Agriculture of the County of Kent*, 1796.
Bracton [Henricus de] *De Legibus et Consuetudinibus Angliae: On the Laws and Customs of England* ed. G.E. Woodbine and S.E. Thorne, 4 vols, Cambridge, Mass., 1968–77.
Brandon, P.F. 'Arable Farming in a Sussex Scarp-foot Parish during the late Middle Ages', *Sussex Archaeological Collections*, c, 1962.
—— 'Kent and Sussex Downland and Coastal Plain' in Finberg and Thirsk, q.v., vol. ii.
Braudel, F. *Civilisation and Capitalism: 15th to 18th Century*, 3 vols, 1981–4.
—— *The Identity of France*, 2 vols, 1988–90.
Bravender, J. 'Farming of Gloucestershire', *Journal of the Royal Agricultural Society of England*, xi, 1850.
Bridgman, C.G.O. (ed.) 'The Burton Abbey Twelfth Century Surveys', *Collections for the History of Staffordshire*, 3rd series, 1918 (1916).
Brookes, C.C. *A History of Steeple Aston and Middle Aston*, Shipston-on-Stour, 1929.
Brooks, F.W. 'Supplementary Stiffkey Papers', in *Camden Miscellany xvi*, Royal Historical Society, Camden 3rd series, lii, 1936.
Brown, W. (ed.) *Yorkshire Deeds*, 3 vols, Yorkshire Archaeological Society Record Series, xxxix, l, lxiii, 1909, 1914, 1922.
—— (ed.) *Yorkshire Star Chamber Proceedings*, Yorkshire Archaeological Society Record Series, xli, 1909.
Brownlow, R. *Reports (A Second Part) of Diverse Famous Cases in Law*, 1652 and various editions.
Bruford, W. H. *Germany in the Eighteenth Century: the social background of the literary revival*, Cambridge, 1959.
Buchanan, R.H. 'Field Systems of Ireland', in Baker and Butlin, q.v.
Burrell, E.D.R. 'An Historical Geography of the Sandlings of Suffolk, 1600–1850', typescript thesis, M.Sc., London University, 1960.
Bushby, D. and Le Hardy, W. *Wormley in Hertfordshire*, 1954.
Butlin, R.A. 'Field Systems of Northumberland and Durham', in Baker and Butlin, q.v.

Caius, J. *The Annals of Gonville and Caius College*, ed. J.A. Venn, Cambridge Antiquarian Society, Octavo Publication, xl, 1904.
Campbell, A. (ed.) *The Charters of Rochester*, British Academy, 1973.
Campbell, B.S.M. 'Agricultural Progress in Medieval England: some evidence from East Norfolk', *Economic History Review*, 2nd series, xxxvi, 1983.
—— 'Population Change and the Generation of Common Fields in East Anglia', *Economic History Review*, 2nd series, xxxiii, 1980.
—— 'The Regional Uniqueness of English Field Systems', *Agricultural History Review*, xxix, 1981.
Cannan, E. *The History of Local Rates in England*, 1896.

Carew, R. *The Survey of Cornwall* (1602), 1769.
Carr, J.P. 'Open Field Agriculture in Mid-Derbyshire', *Derbyshire Archaeological Journal*, lxxxiii, 1963.
Carsten, F.L. 'The Slavs in Northeast Germany', *Economic History Review*, xi, 1941.
Carus-Wilson, E.M. (ed.) *Essays in Economic History*, 3 vols, 1954–62.
Chambers, J.D. and Mingay, G.E. *The Agricultural Revolution 1750–1880*, 1966.
Chandler, H.W. (ed.) *Five Court Rolls of Great Cressingham in the County of Norfolk*, 1885.
Chapman, V. 'Open Fields in West Cheshire', *Transactions of the Historic Society of Lancashire and Cheshire*, civ, 1953.
Charles-Edwards, T.M. 'Kinship, Status and the Origins of the Hide', *Past and Present* no. 56, 1972.
Chayanov, A.V. *The Theory of Peasant Economy*, ed. D. Thorner, B. Kerblay and R.E.F. Smith, Homewood, Ill., 1966.
Chibnall, A.C. *Sherington: the fiefs and fields of a Buckinghamshire village*, Cambridge, 1965.
Chibnall, M. (ed.) *Select Documents of the English Lands of the Abbey of Bec*, Royal Historical Society, Camden 3rd series, lxxiii, 1951.
Childrey, J. *Britannia Baconica*, 1660.
Chorley, P. 'English Cloth Exports during the 13th and 14th Centuries', *Historical Research*, lxi, 1988.
Christians, C. 'Quelques aspects de la structure agraire et de l'aménagement rural en Ardenne Belge', in *Géographie et Histoire Agraires*, Annales de l'Est, q.v.
Clapham, J.H. *The Economic Development of France and Germany 1815–1914*, Cambridge, 1936.
Clarke, W.G. *In Breckland Wilds*, 1926.
Clay, C.T. (ed.) *Yorkshire Deeds iv and v*, Yorkshire Archaeological Society Record Series, lxv, lxix, 1924, 1926.
Cobbett, W. *Rural Rides*, Everyman edition, 2 vols.
Coleman, D.C. 'The Economy of Kent under the later Stuarts', typescript thesis, Ph.D., London University, 1951.
Collier, C.V. (ed.) 'Stovin's Manuscript', *Transactions of the East Riding Archaeological Society*, xii, 1905 (1904).
Colman, F.S. *A History of Barwick-in-Elmet*, Leeds, 1908.
Colvin, H.M. *A History of Deddington, Oxfordshire*, 1963.
Cooke, A.H. *The Early History of Mapledurham*, Oxfordshire Record Society, vii, 1925.
Cooper, G.M. 'Berwick Parochial Records', *Sussex Archaeological Collections*, vi, 1853.
Cooper, W. *Wootton Waven: its History and Records*, Leeds, 1936.
Coopland, G.W. *The Abbey of Saint-Bertin and its Neighbourhood, 900–1350*, in Oxford Studies in Social and Legal History, vol. iv, Oxford, 1914.
Corbett, W.J. 'Elizabethan Village Surveys', *Transactions of the Royal Historical Society*, new series, xi, 1897.
Cornwall, J.C.K. 'The Agrarian History of Sussex, 1560–1640', typescript thesis, M.A., London University, 1953.

—— 'Farming in Sussex, 1560–1640', *Sussex Archaeological Collections*, xcii, 1954.

Coulton, G. G. *The Medieval Village*, Cambridge, 1925.

Crofton, H. T. 'Relics of the Common Field System in and near Manchester', *Manchester Quarterly*, January 1887.

Cromarty, D. *The Fields of Saffron Walden in 1400*, Chelmsford, 1966.

Crossley, E. W. 'The Testamentary Documents of Yorkshire Peculiars', in *Miscellanea vol. ii*, Yorkshire Archaeological Society Record Series, lxxiv, 1929.

Crutchley, J. *A General View of the Agriculture of the County of Rutland*, 1794.

Cunningham, W. 'Common Rights at Cottenham and Stretham in Cambridgeshire', in *Camden Miscellany xii*, Royal Historical Society, Camden 3rd series, xviii, 1910.

Curwen, E. C. 'Ancient Cultivations at Grassington, Yorkshire', *Antiquity*, ii, 1928.

—— *Plough and Pasture*, 1946.

Dahlman, C. J. *The Open-Field System and Beyond: a property rights analysis of an economic institution*, Cambridge, 1980.

Darbyshire, H. S. and Lumb, G. D. *The History of Methley*, Publications of the Thoresby Society, xxv, 1937.

Darlington, R. R. (ed.) *The Cartulary of Darley Abbey*, 2 vols, p.p. 1945.

—— (ed.) *The Cartulary of Worcester Cathedral Priory, Register I*, Pipe Roll Society, new series, lxxvi, 1968 (1962–3).

Davies, M. 'Common Lands in Southeast Monmouthshire', *Cardiff Naturalists' Society Reports and Transactions*, lxxxv, 1955–6.

—— 'Field Patterns in the Vale of Glamorgan', *Cardiff Naturalists' Society Reports and Transactions*, lxxxiv, 1954–5.

—— 'Rhosili Open Field and related South Wales Field Patterns', *Agricultural History Review*, iv, 1956.

—— 'The Field Systems of South Wales', in Baker and Butlin, q.v.

—— 'The Open Fields of Laugharne', *Geography*, xl, 1955.

Davies, W. *A General View of the Agriculture and Domestic Economy of South Wales*, 2 vols, 1814.

Davis, R. C. H. 'East Anglia and the Danelaw', *Transactions of the Royal Historical Society*, 5th series, v, 1955.

Davis, T. *A General View of the Agriculture of the County of Wiltshire*, 1794, 1813.

Davison, A. J. 'Some Aspects of the Agrarian History of Hougham and Snetterton as revealed in the Buxton MSS', *Norfolk Archaeology*, xxxv, 1973.

Day, J. 'Peuplement, cultures et régimes fonciers en Trexenta (Sardaigne) XIIIe–XIXe siècle', in Guarducci, q.v.

Defoe, D. *A Tour through England and Wales*, Everyman edition, 2 vols.

Defourneaux, M. 'Le Problème de la terre en Andalousie au XVIIIe siècle et les projets de réforme agraire', *Revue Historique*, ccxvii, 1957.

D'Elboux, R. H. (ed.) *Surveys of the Manors of Robertsbridge, Sussex, and Michelmarsh, Hampshire, and of the Demesne Lands of Halden in Rolvenden, Kent, 1567–70*, Sussex Record Society, xlvii, 1944.

Delisle, L. *Etudes sur la condition de la classe agricole et l'état de l'agriculture en Normandie au Moyen Age*, Paris, 1903.
De Roover, R. (ed.) *Gresham on Foreign Exchange*, Cambridge, Mass., 1949.
De Vries, J. *Dutch Rural Economy in the Golden Age 1500–1700*, New Haven and London, 1974.
Dilley, R. S. 'The Cumberland Court Leet and the Use of Common Lands', *Transactions of the Cumberland and Westmorland Antiquarian and Archaeological Society*, new series, lxvii, 1967.
Dobenecker, O. (ed.) *Regesta Diplomatica necnon Epistolaria Historiae Thuringiae*, vol. i (*c.* 500–1152), Jena, 1896.
Dodd, K. M. (ed.) *The Field Book of Walsham-le-Willows, 1577*, Suffolk Records Society, xvii, 1974.
Dodgshon, R. A. *The Origins of British Field Systems: an interpretation*, London and New York, 1980.
Donaldson, J. *A General View of the Agriculture of the County of Northampton*, Edinburgh, 1794.
Donnithorne, A. *China's Economic System*, 1967.
Dopsch, A. *The Economic and Social Foundation of European Civilization*, 1937.
Dorošenko, V. 'Der ostbaltische Herrenhof des 16.–18. Jh. als "Getreidefabrik"' in Guarducci, q.v.
Douch, R. 'Customs and Traditions of the Isle of Portland', *Antiquity*, xxiii, 1949.
Douglas, D. C. *The Social Structure of Medieval East Anglia*, Oxford Studies in Social and Legal History, ix, Oxford, 1927.
Duby, G. *L'Economie rurale et la vie des campagnes dans l'occident médiéval: France, Angleterre, Empire IXe–XVe siècle*, 2 vols, Paris, 1962.
—— 'La Révolution agricole médiévale', *Revue de Géographie de Lyon*, xxix, 1954.
Dyer, C. *Lords and Peasants in a Changing Society: the estates of the Bishopric of Worcester, 680–1540*, Cambridge, 1980.
—— 'The West Midlands' in Finberg and Thirsk, q.v., vol. ii.

Eden, F. M. *The State of the Poor*, 3 vols, 1797.
Eland, G. *At the Courts of Great Canfield, Essex*, 1949.
—— (ed.) *Purefoy Letters 1735–53*, 2 vols, 1931.
Elgee, F. *The Moorlands of Northeast Yorkshire*, 1912.
Elliott, G. (G.) 'The Enclosure of Aspatria', *Transactions of the Cumberland and Westmorland Antiquarian and Archaeological Society*, new series, lx, 1960.
—— 'The Field Systems of Northwest England', in Baker and Butlin, q.v.
—— 'The System of Cultivation ... in Cumberland Open Fields', *Transactions of the Cumberland and Westmorland Antiquarian and Archaeological Society*, new series, lix, 1959.
—— 'The System of Cultivation and the Evidence of Enclosure in Cumberland Open Fields in the Sixteenth Century', in *Géographie et Histoire Agraires*, Annales de l'Est, q.v.
Elliott, S. 'The Open Field System of an Urban Community: Stamford in the Nineteenth Century', *Agricultural History Review*, xx, 1972.
Ellis, W. *A Compleat System of Experienced Improvements made on Sheep, Grass-lambs and House-lambs*, 1749.

—— *Chiltern and Vale Farming Explained*, [1733].
—— *The Modern Husbandman*, 8 vols, 1740.
Emerson, W. R. 'The Economic Development of the Estates of the Petre Family in Essex in the Sixteenth and Seventeenth Centuries', typescript thesis, D.Phil., Oxford University, 1951.
Emmison, F. G. (ed.) *Jacobean Household Inventories*, Bedfordshire Historical Record Society, xx, 1938.
—— *Types of Open Field Parishes in the Midlands*, Historical Association Pamphlet 108, 1937.
Ernle (Lord) *English Farming Past and Present*, 1927.
Evans, E. J. and Beckett, J. V. 'Cumberland, Westmorland and Furness', in Finberg and Thirsk, q.v., v, pt i.
Evans, I. L. *The Agrarian Revolution in Roumania*, Cambridge, 1924.
Evans, N. 'The Community of South Elmham, Suffolk, 1550–1640', typescript thesis, M.Phil., University of East Anglia, 1978.
Everitt, A. *Continuity and Colonization: the evolution of Kentish settlement*, Leicester, 1986.

Farey, J. *A General View of the Agriculture and Minerals of Derbyshire*, 3 vols, 1811–17.
Farnham, G. F. *Quorndon Records*, sine loco, 1912.
Fei, Hsiao-Tung, *Peasant Life in China*, 1939.
Festy, O. *L'Agriculture pendant la Révolution Française*, 2 vols, Paris, 1947–50.
Finberg, H. P. R. *Early Charters of Wessex*, Leicester, 1964.
—— *Lucerna*, 1964.
—— *Tavistock Abbey*, Cambridge, 1951.
—— 'The Open Field in Devonshire', *Antiquity*, xxiii, 1949.
Finberg, H. P. R. and Thirsk, J. (gen. eds) *The Agrarian History of England and Wales*, Cambridge, several volumes, 1967– in progress.
Fisher, F. N. 'Notes on the Manorial History of Horsley', *Derbyshire Archaeological Journal (Jnl Derbys. Archaeol. and Nat. Hist. Soc.)*, xxv, 1952.
Fishwick, H. (ed.) *The Survey of the Manor of Rochdale 1626*, Chetham Society, new series, lxxi, 1913.
Fitzherbert, *The Boke of Husbandrie*, 1523.
—— *The Book of Husbandry*, ed. W. W. Skeat, English Dialect Society, xxvii, 1882.
—— *The Boke of Surveyinge and Improvementes*, 1535.
Fletcher, W. G. *History of Loughborough*, Loughborough, 1887.
Flower, R. and Smith, H. (eds) *The Parker Chronicle and Laws: Corpus Christi College Cambridge MS 173, a facsimile*, Early English Text Society, 1941.
Folkingham, W. *Feudigraphia: the synopsis or epitome of surveying*, 1610.
Fortrey, S. *England's Interest and Improvement*, Cambridge, 1663.
Fourquin, G. *Les Campagnes de la région Parisienne à la fin du Moyen Age, du milieu du XIIIe siècle au début du XVIe siècle*, Paris, 1964.
Fox, H. S. A. 'Approaches to the Adoption of the Midland System', in T. Rowley, q.v.
—— 'The Chronology of Enclosure and Economic Development in Medieval Devon', *Economic History Review*, 2nd series, xxviii, 1975.

—— 'Field Systems of East and South Devon. Part I: East Devon', *Report and Transactions of the Devonshire Association*, civ, 1972.
Franz, G. *Geschichte des deutschen Bauernstandes vom frühen Mittelalter bis zum 19. Jahrhundert*, Stuttgart, 1970.
Fream, W. *Elements of Agriculture*, 1918.
French, R. A. 'The Three-Field System of Sixteenth-Century Lithuania', *Agricultural History Review*, xviii, 1970.
Fry, G. S. and E. A. (eds) *Abstracts of Wiltshire Inquisitiones Post Mortem, temp. Car. I*, Index Library, British Record Society, 1901.
Fussell, G. E. 'Farming Systems in the Classical Era', *Technology and Culture*, viii, 1967.
—— (ed.) *Robert Loder's Farm Accounts 1610–1620*, Royal Historical Society, Camden 3rd series, liii, 1936.

Gerarde, J. *The Herball*, 1597.
Giles, J. A. *History of the Parish and Town of Brampton*, Brampton, 1848.
Gill, H. and Guilford, E. L. (eds) *The Rector's Book of Clayworth 1672–1701*, Nottingham, 1910.
Gissel, S. 'Village and Environment. A Case: Danish Isles in the later Middle Ages', in Guarducci, q.v.
Goddard, C. V. (ed.) 'Customs of the Manor of Winterbourne Stoke 1574', *Wiltshire Archaeological Magazine*, xxxiv, 1905–6.
Godfrey, W. H. (ed.) *The Booke of John Rowe*, Sussex Record Society, xxxiv, 1928.
Gonner, E. C. K. *Common Land and Inclosure*, 1912.
Gooch, W. *A General View of the Agriculture of the County of Cambridge*, 1813.
Goodman, F. R. *Reverend Landlords and their Tenants*, Winchester, 1930.
Graham, T. H. B. (ed.) *The Barony of Gilsland: Lord William Howard's Survey taken in 1603*, Cumberland and Westmorland Antiquarian and Archaeological Society, extra series, xvi, 1934.
Gras, N. S. B. *The Early English Customs System*, Cambridge, Mass., 1918.
Gras, N. S. B. and E. C. *The Economic and Social History of an English Village*, Cambridge, Mass., 1930.
Gray, H. L. *English Field Systems*, Cambridge, Mass., 1915.
Gray, M. 'The Regions and their Issues: Scotland', in G. E. Mingay, q.v.
Greig, W. *General Report on the Gosford Estates in County Armagh*, ed. F. M. L. Thompson and D. Tierney, Belfast, 1976.
Gretton, R. H. (ed.) *The Burford Records*, Oxford, 1920.
Guarducci, A. (ed.) *Agricoltura e Trasformazione dell'Ambiente secoli XIII–XVIII*, Istituto Internazionale di Storia Economica 'Francesco Datini', Prato, 1984.
Gulley, J. L. M. 'The Wealden Landscape in the Early Seventeenth Century and its Antecedents', typescript thesis, Ph.D., London University, 1960.

Hale, W. H. (ed.) *Registrum sive Liber Irrotularius et Consuetudinarius Prioratus Beatae Mariae Wigorniensis*, Camden Society, xci, 1865.
—— (ed.) *The Domesday of St Pauls in the Year MCCXXII or Registrum de Visitatione Maneriorum per Robertum Decanum*, Camden Society, lxix, 1858.

Hall, H. *Society in the Elizabethan Age*, 1887.
Hall, R. de Z. 'Post-Medieval Land Tenure, Preston Plucknett', *Report of the Somersetshire Archaeological and Natural History Society*, cv, 1961.
Hallam, H. E. 'Eastern England' and 'Southern England', in Finberg and Thirsk, q.v., ii.
—— *Settlement and Society: a study of the early agrarian history of South Lincolnshire*, Cambridge, 1965.
Halliwell, J. O. (ed.) *Ancient Inventories of Furniture, Pictures, Tapestry, Plate etc. Illustrative of the Domestic Manners of the English in the Sixteenth and Seventeenth Centuries*, 1854.
Hammond, R. J. 'Social and Economic Circumstances of Ket's Rebellion', typescript thesis, M.A., London University, 1933.
Handley, J. E. *Scottish Farming in the Eighteenth Century*, 1953.
Hanssen, G. *Agrarhistorische Abhandlungen*, 2 vols, Leipzig, 1880–4.
Hard, G. 'Ein Vierzelgensystem des 18. Jh', *Zeitschrift für Agrargeschichte und Agrarsoziologie*, xi, 1963.
Harmer, F. E. (ed.) *Select English Historical Documents of the Ninth and Tenth Centuries*, Cambridge, 1914.
Harris, A. *The Open Fields of East Yorkshire*, East Yorkshire Local History Society, 1974.
—— *The Rural Landscape of the East Riding of Yorkshire 1700–1850*, 1961.
Harris, M. D. (ed.) *The Coventry Leet Book 1420–1555*, Early English Text Society, cxxxiv, cxxxv, cxxxviii, cxlvi, 1907–13.
Harrison, J. *An Economic History of Modern Spain*, Manchester, 1978.
Hart, C. R. *The Early Charters of Eastern England*, Leicester, 1966.
Hart, W. H. (ed.) *Historia et Cartularium Monasterii Sancti Petri Gloucestriae*, 3 vols, Rolls Series, 1863–7.
Hartley, M. and Ingilby, J. *Yorkshire Village*, 1953.
Harvey, L. A. and St Leger-Gordon, D. *Dartmoor*, 1953.
Harvey, P. D. A. *An Oxfordshire Medieval Village: Cuxham, 1240–1400*, 1965.
—— 'The English Inflation 1180–1220', in R. H. Hilton, *Peasants, Knights and Heretics*, q.v.
Hassall, C. *A General View of the Agriculture of the County of Carmarthen*, 1794.
Hassall, W. O. (ed.) *Wheatley Records 956–1956*, Oxfordshire Record Society, 1956.
Hatcher, J. *Rural Economy and Society in the Duchy of Cornwall 1300–1500*, Cambridge, 1970.
—— 'Southwest England', in Finberg and Thirsk, q.v., ii.
Haudricourt, A. G. and Delamarre, M. J.-B. *L'Homme et la charrue à travers le monde*, Paris, 1955.
Havinden, M. A. 'Agricultural Progress in Open-Field Oxfordshire', *Agricultural History Review*, ix, 1961, reprinted in E. L. Jones, q.v.
—— (ed.) *Household and Farm Inventories in Oxfordshire 1550–1590*, Historical MSS Commission Joint Publication 10 and Oxfordshire Record Society xliv, 1965.
—— 'The Rural Economy of Oxfordshire, 1580–1730', typescript thesis, B.Litt., Oxford University, 1961.
Heckscher, E. *An Economic History of Sweden*, Cambridge, Mass., 1963.

[Hervey, Lord Francis] (ed.) *Ickworth Survey Booke*, sine loco nec data.

Hey, D.G. *An English Rural Community: Myddle under the Tudors and Stuarts*, Leicester, 1974.

Higounet, C. 'L'Assolement triennal dans la plaine de France au XIIIe siècle', *Comptes Rendus des Séances de l'Année 1956*, Académie des Inscriptions et Belles-Lettres, Paris, 1956.

—— *Structure et exploitation d'un terroir Cistercien de la plaine de France XIIe–XVe siècle*, Ecole Pratique des Hautes Etudes, VIe section: Centre de Recherches Historiques: Les Hommes et la Terre, no. X, Paris, 1965.

Hill, M.C. 'The Wealdmoors, 1560–1660', *Transactions of the Shropshire Archaeological Society*, liv, 1953 (1951–3).

Hilton, R.H. *A Medieval Society: the West Midlands at the End of the Thirteenth Century*, 1967.

—— (ed.) *Peasants, Knights and Heretics: studies in medieval English social history*, Cambridge, 1981.

—— *The Economic Development of some Leicestershire Estates in the 14th and 15th Centuries*, Oxford, 1947.

Hine, R.L. *The History of Hitchin*, 2 vols, 1927.

Historical Manuscripts Commission, *Manuscripts of Lord Middleton preserved at Wollaton Hall, Notts.*, H.M.S.O., Cd.5567, 1911.

Hoare, C.M. *The History of an East Anglian Soke*, Bedford, 1918.

Hoare, R.C. *The History of Modern Wiltshire*, 6 vols, 1822–44.

Hoebaux, J.J. 'Encore et toujours à propos de l'assolement triennal au Moyen Age', *Revue Belge de Philologie et d'Histoire / Belgisch Tijdschrift voor Filologie en Geschiedenis*, xlii, 1966.

Hogan, M.P. 'Clays, *Culturae*, and the Cultivator's Wisdom', *Agricultural History Review*, xxxvi, 1988.

Holderness, B.A. 'East Anglia and the Fens', in Finberg and Thirsk, q.v., v, pt i.

Holinshed, R. and Harrison, W. *Chronicles*, vols i and ii, 1586.

Holland, H. *A General View of the Agriculture of Cheshire*, 1808.

Hollings, M. (ed.) *The Red Book of Worcester*, Worcestershire Historical Society, 4 pts, 1934–50.

Homans, G.C. *English Villagers of the Thirteenth Century*, Cambridge, Mass., 1941.

Hömberg, A. *Die Entstehung der westdeutschen Flurformen: Blockgemengflur, Streifenflur, Gewannflur*, Berlin, 1935.

—— *Siedlungsgeschichte des oberen Sauerlandes*, Geschichtliche Arbeiten zur westfälischen Landesforschung, Band 3, Münster, 1938.

Hone, N.J. *The Manor and Manorial Records*, Antiquaries Books, 1912.

Hooke, D. 'Open-field Agriculture – the Evidence from Pre-Conquest Charters of the West Midlands', in R. Rowley, q.v.

Hoskins, W.G. *Essays in Leicestershire History*, Liverpool, 1950.

—— 'The Fields of Wigston Magna', *Transactions of the Leicestershire Archaeological Society*, xix, 1936–7.

—— *The Midland Peasant*, 1957.

Hoskins, W.G. and Finberg, H.P.R. *Devonshire Studies*, 1952.

Hövermann, J. *Die Entwicklung der Siedlungsformen in den Marschen des Elbe-Wesel-Winkels*, Forschungen zur deutschen Landeskunde, Band 56, Remagen, 1951.

Howard, C. 'On the Culture and Planting of Saffron', in *Philosophical Transactions of the Royal Society of London*, ed. C. Hutton, G. Shaw and R. Pearson, 18 vols, 1809, ii.

Howson, M. F. 'Aughton, near Lancaster: the story of a northwestern hamlet through the centuries', *Transactions of the Lancashire and Cheshire Antiquarian Society*, lxix, 1959.

Hudson, W. 'Traces of Primitive Agricultural Organisation as suggested by a Survey of the Manor of Martham, Norfolk (1101–1292), *Transactions of the Royal Historical Society*, 4th series, i, 1918.

Hughes, E. *North Country Life in the Eighteenth Century:* [*i*] *The North-East 1700–50*, 1952.

Hull, F. 'Agriculture and Rural Society in Essex, 1560–1640', typescript thesis, Ph.D., London University, 1950.

Hulton, W.A. (ed.) *The Coucher Book or Cartulary of Whalley Abbey*, 3 vols, Chetham Society, x, xi, xvi, 1863–7.

Ilešič, S. *Die Flurformen Sloweniens im Lichte der europäischen Flurforschung*, being Münchner Geographische Hefte, Heft 16, Materialien zur Agrargeographie V, 1959.

—— 'Die Jüngeren Gewannfluren in Nordwestjugoslawien', *Geografiska Annaler*, xliii, 1961.

—— 'Les Problèmes du paysage rural en Yougoslavie nordoccidentale, et specialement en Slovenie' in *Géographie et Histoire Agraires*, Annales de l'Est, q.v.

Irminon, *Polyptyque de l'Abbé Irminon, ou dénombrement des manses, des serfs et des revenus de l'abbaye de Saint-Germain-des-Prés*, ed. B. E. C. Guérard, 2 vols, Paris, 1844.

Irsigler, F. 'Intensivwirtschaft, Sonderkulturen und Gartenbau als Elemente der Kulturlandschaftgestaltung in den Rheinlanden (13.–16. Jahrhundert)' in Guarducci, q.v.

Jackson, C. 'Notes from the Court Rolls of the Manor of Epworth in the Co. of Lincoln', *The Reliquary*, xxiii, 1882–3.

Jackson, J. C. 'Open Field Cultivation in Derbyshire', *Derbyshire Archaeological Journal*, lxxxii, 1962.

Jackson, J. E. (ed.) *Liber Henrici de Soliaco Abbatis Glaston. et vocatur A: An Inquisition of the Manors of Glastonbury Abbey in the year MCLXXXIX*, Roxburghe Club, 1882.

Jacquart, J. 'French Agriculture in the Seventeenth Century' in P. Earle (ed.) *Essays in European Economic History 1500–1800*, Oxford, 1974.

—— *La Crise rurale en Ile-de-France 1550–1670*, Publications de la Sorbonne, N.S., Recherches no. 10, Paris, 1974.

—— 'Le Rôle de la grande exploitation dans la formation des paysages des plaines Limoneuses de la France du Nord (c. 1450–c. 1800)' in Guarducci, q.v.

James, W. and Malcolm, J. *A General View of the Agriculture of the County of Buckingham*, 1794.

—— *A General View of the Agriculture of the County of Surrey*, 1794.

Jeffery, R. W. *The Manors and Advowson of Great Rollright*, Oxfordshire Record Society, 1927.
Jenkins, J. G. (ed.) *Cartulary of Missenden Abbey, pt i*, Buckinghamshire Archaeological Society Records Branch, ii, 1938.
Jiriček, C. *Geschichte der Serben*, 2 vols, Gotha, 1911–18.
John, E. *Land Tenure in Early England*, Leicester, 1960.
Johnsen, O. A. *Norwegische Wirtschaftsgeschichte*, Jena, 1939.
Jolliffe, J. E. A. *Pre-Feudal England: the Jutes*, 1933.
Jones, E. L. (ed.) *Agriculture and Economic Growth in England 1650–1815*, 1967.
Jones, G. R. J. 'Early Customary Tenures in Wales and Open Field Agriculture' in T. Rowley, q.v.
—— 'The Field Systems of North Wales' in Baker and Butlin, q.v.
—— 'Medieval Open Fields and associated Settlement Patterns in North-west Wales' in *Géographie et Histoire Agraires*, Annales de l'Est, q.v.
Juillard, E. 'L'Assolement biennal dans l'agriculture septentrionale: le cas particulier de la Basse-Alsace, *Annales de Géographie*, lxi, 1952.
—— *La Vie rurale dans la plaine de Basse-Alsace*, Paris, 1953.
Juillard, E., Meynier, A., Planhol, X. de and Sautter, G. *Structures Agraires et Paysages Ruraux*, Annales de l'Est, q.v.
Jutikkala, E. 'How Open Fields came to be divided into numerous Selions', *Proceedings of the Finnish Academy of Science and Letters 1952*, Helsinki (Helsingfors), 1953.

Kalm, P. *Kalm's Account of his Visit to England*, 1892.
Kay, G. *A General View of the Agriculture of North Wales*, Edinburgh, 1794.
Keating, G. *Rural Economy in the Bombay Deccan*, 1912.
Kemble, J. M. (ed.) *Codex Diplomaticus Aevi Saxonici*, 6 vols, 1839–48.
Kemp, B. R. (ed.) *Reading Abbey Cartularies*, 2 vols, Royal Historical Society, Camden 4th series, xxxi, xxxiii, 1986–7.
Kerridge, E. *Agrarian Problems in the Sixteenth Century and After*, 1969.
—— 'A Reconsideration of some former Husbandry Practices', *Agricultural History Review*, iii, 1955.
—— *The Agricultural Revolution*, 1967.
—— *The Farmers of Old England*, 1973.
—— 'Ridge and Furrow and Agrarian History', *Economic History Review*, 2nd series, iv, 1951.
—— (ed.) *Surveys of the Manors of Philip, Earl of Pembroke and Montgomery 1631–2*, Wiltshire Records Society (Wilts. Archaeol. and Nat. Hist. Soc. Recs Brch), ix, 1953.
—— *Textile Manufactures in Early Modern England*, Manchester, 1985.
King, E. (ed.) 'Estate Records of the Hotot Family', in E. King (ed.) *A Northamptonshire Miscellany*, Northamptonshire Record Society, xxxii, 1983 (1982).
King, E. J. *Years Beyond Memory*, Oxford, 1954.
Kingston, A. *A History of Royston*, 1906.
Kirbis, W. *Siedlungs- und Flurformen germanischer Länder, besonders Grossbritanniens, im Lichte der deutschen Siedlungsforschung*, Göttinger Geographische Abhandlungen, Heft 10, 1952.

Kitchin, G.W. *The Manor of Manydown*, Hampshire Record Society, x, 1895.
Klag, E. 'Ein Beitrag zur Zweifelderwirtschaft', *Zeitschrift für Agrargeschichte und Agrarsoziologie*, xvii, 1969.
Knappen, M.M. *Tudor Puritanism: a chapter in the history of idealism*, Chicago and London, 1970.
Knocker, H.W. 'The Evolution of Holmesdale: no. 3, the manor of Sundrish', *Archaeologia Cantiana*, xliv, 1932.
Kovalevsky, M. *L'Agriculture en Russie*, extrait de la *Revue Internationale de Sociologie*, Paris, 1897.
—— *Modern Customs and Ancient Laws of Russia*, 1891.
Krenzlin, A. 'Blockflur, Langstreifenflur und Gewannflur als Ausdruck agrarischer Wirtschaftsformen in Deutschland', in *Géographie et Histoire Agraires*, Annales de l'Est, q.v.
—— 'Probleme der neueren nordostdeutschen und ostmitteldeutschen Flurformenforschung', *Deutsches Archiv für Landes- und Volksforschung*, iv, 1940.
—— 'Zur Genese der Gewannflur in Deutschland nach Untersuchungen im nördlichen Unterfranken', *Geografiska Annaler*, xliii, 1961.
Krenzlin, A. and Reusch, L. *Die Entstehung der Gewannflur nach Untersuchungen im nördlichen Unterfranken*, Frankfurter Geographische Hefte, xxxv, 1. Heft, Frankfurt-am-Main, 1961.
Krzymowski, R. *Die landwirtschaftlichen Wirtschaftssysteme Elsass-Lothringens*, Gebweiler im Elsass, 1914.

Lambert, H.C.M. *History of Banstead in Surrey*, 2 vols, Oxford, 1912.
Lamprecht, K. *Deutsches Wirtschaftsleben im Mittelalter*, Leipzig, 1886.
Laslett, P. *Household and Family in Past Time*, Cambridge, 1972.
Latouche, R. *The Birth of Western Economy*, 1961.
Laur, E. *Swiss Farming*, Berne, 1949.
Laurence, E. *The Duty of a Steward to his Lord*, 1727.
Laurence, J. *A New System of Agriculture*, 1726.
Lavelye, E. de, 'The Land System of Holland and Belgium', in Probyn, q.v.
Lawson-Tancred, T. 'Three Seventeenth-Century Court Rolls of the Manor of Aldborough', *Yorkshire Archaeological Journal*, xxxv, 1943.
Leadam, I.S. (ed.) *The Domesday of Inclosures*, 2 vols, 1897.
Leconfield (Lord) *Petworth Manor in the Seventeenth Century*, 1954.
—— *Sutton and Duncton Manors*, 1956.
Lefèbvre, G. *Les Paysans du nord pendant la Révolution Française*, Bari, 1959.
Leister, I. 'Zum Probleme des "Keltischen Einzelhofs" in Irland', in *Géographie et Histoire Agraires*, Annales de l'Est, q.v.
Leland, J. *The Itinerary*, ed. L.T. Smith, 5 vols, 1906–10.
Lennard, R. (V.) 'English Agriculture under Charles II: the evidence of the Royal Society's "Enquiries" ', *Economic History Review*, iv, 1932–4, reprinted in W.E. Minchinton, q.v.
Leonard, E.M. *The Early History of English Poor Relief*, Cambridge, 1900.
Levett, A.E. *Studies in Manorial History*, Oxford, 1938.
Lewinski, J. St. *The Origin of Property and the Formation of the Village Community*, 1913.
Lewis, A.R. *The Northern Seas: shipping and commerce in Northern Europe A.D. 300–1100*, New York, 1978.

Lewis, T. 'Seebohm's Tribal System of Wales', *Economic History Review*, 2nd series, ix, 1956.

Lindley, E. S. 'The History of Wortley in the Parish of Wotton-under-Edge, *Transactions of the Bristol and Gloucestershire Archaeological Society*, lxviii, lxix, 1949–50.

Lisle, E. *Observations in Husbandry*, 1757.

Livet, R. 'Les Champs allongés de Basse-Provence', in *Géographie et Histoire Agraires*, Annales de l'Est, q.v.

Lloyd, J. E. *History of Wales*, 2 vols, 1939.

Lobel, M. D. *The Borough of Bury St Edmunds*, Oxford, 1935.

Loewe, K. von, 'Commerce and Agriculture in Lithuania 1400–1600', *Economic History Review*, 2nd series, xxvi, 1973.

Lomas, H. A. *History of Abbotsham*, Abbotsham, 1956.

Longman, G. *A Corner of England's Garden: an agrarian history of southwest Hertfordshire 1600–1850*, 2 vols, sine loco nec data.

Lot, F. *The End of the Ancient World and the Beginnings of the Middle Ages*, 1931.

Lütge, F. *Die Agrarverfassung des frühen Mittelalters im mitteldeutschen Raum, vornahmlich in der Karolingerzeit*, Jena, 1937.

—— *Geschichte der deutschen Agrarverfassung vom frühen Mittelalter bis zum 19. Jahrhundert*, Stuttgart, 1967.

Lyaschenko, P. I. *History of the National Economy of Russia to the 1917 Revolution*, New York, 1949.

McCourt, D. 'Infield and Outfield in Ireland', *Economic History Review*, 2nd series, vii, 1955.

McGurk, P., Dumville, D. M., Godden, M. R. and Knock, A. (eds) *An Eleventh Century Anglo-Saxon Illustrated Miscellany* (Early English Manuscripts in Facsimile, ed. G. Harlow, no. 21), Copenhagen, 1983.

McKisack, M. *The Fourteenth Century: 1307–1399*, Oxford, 1961.

MacLear, A. B. *Early New England Towns*, New York, 1908.

Macnab, J. W. 'British Strip Lynchets', *Antiquity*, xxxix, 1965.

Madox, F. (ed.) *Formulare Anglicanum, or a Collection of Ancient Charters and Instruments*, 1702.

Maitland, F. W. *Domesday Book and Beyond: three essays in the early history of England*, Cambridge, 1897.

Manley, F. 'Customs of the Manor of Purton', *Wiltshire Archaeological Magazine*, xl, 1917–19.

Markham, G. *The English Husbandman*, 1635.

—— *The Inrichment of the Weald of Kent*, 1625.

Marshall, W. *A General View of the Agriculture of the Central Highlands of Scotland*, 1794.

—— *A Review of the Reports to the Board of Agriculture from the Eastern Department of England*, 1811.

—— *A Review (and Complete Abstract) of the Reports to the Board of Agriculture from the Midland Department of England*, 1815.

—— *A Review of the Reports to the Board of Agriculture from the Northern Department of England*, 1808.

—— *A Review (and Complete Abstract) of the Reports to the Board of Agriculture from the Southern and Peninsular Departments of England*, 1817.
—— *A Review of the Reports to the Board of Agriculture from the Western Department of England*, 1810.
—— *Experiments and Observations concerning Agriculture and the Weather*, 1779.
—— *Minutes, Experiments, Observations and General Remarks on Agriculture in the Southern Counties*, 2 vols, 1799.
—— *On the Appropriation and Inclosure of Commonable and Intermixed Lands*, 1801.
—— *On the Landed Property of England*, 1804.
—— *The Rural Economy of Glocestershire*, 2 vols, Gloucester, 1789.
—— *The Rural Economy of the Midland Counties*, 2 vols, 1790.
—— *The Rural Economy of Norfolk*, 2 vols, 1787.
—— *The Rural Economy of the Southern Counties*, 2 vols, 1798.
—— *The Rural Economy of the West of England*, 2 vols, 1796.
—— *The Rural Economy of Yorkshire*, 2 vols, 1788.
Maskelyne, T. S. and Manley, F. H. 'Notes on the Ecclesiastical History of Wroughton', *Wiltshire Archaeological Magazine*, xli, 1917–22.
Mason, E. (ed.) *The Beauchamp Cartulary, Charters 1100–1268*, Pipe Roll Society, new series, xliii, 1980 (1971–3).
Mastin, J. *The History and Antiquities of Naseby*, Cambridge, 1792.
Mavor, J. *An Economic History of Russia*, 2 vols, 1925.
Mavor, W. *A General View of the Agriculture of Berkshire*, 1808.
Maxey, E. *A New Inst*[*r*]*uction of Plowing and Setting of Corn*, 1601.
Mayhew, A. *Rural Settlement and Farming in Germany*, 1973.
Mead, W. R. 'Ridge and Furrow in Buckinghamshire', *Geographical Journal*, cxx, 1954.
Meitzen, A. *Siedelung and Agrarwesen der Westgermanen und Ostgermanen, der Kelten, Römer, Finnen und Slawen* (being Pt I of *Wanderungen, Anbau und Agrarrecht der Völker Europas nördlich der Alpen*), 3 vols, Berlin, 1895.
Meyer, A. 'La Structure agraire dans la commune d'Otterswiller', *Annales de Géographie*, xlvi, 1937.
Middleton, J. *View of the Agriculture of Middlesex*, 1798.
Miller, E. 'Northern England', in Finberg and Thirsk, q.v., ii.
Millican, P. *A History of Horstead and Stanninghall*, Norwich, 1937.
Minchinton, W. E. (ed.) *Essays in Agrarian History*, 2 vols, Newton Abbot, 1968.
Mingay, G. E. (ed.) *The Victorian Countryside*, 2 vols, London and Boston, 1981.
Money, W. (ed.) *A Purveyance of the Royal Household in the Elizabethan Age, 1575*, 1901.
Mortensen, H. and G. *Die Besiedlung des nordöstlichen Ostpreussens bis zum Beginn des 17. Jahrhunderts*, 2 pts, being vols 7 and 8 of the 'Deutschland und der Osten' series, Leipzig, 1937–8.
Mortimer, J. *The Whole Art of Husbandry*, 1707.
Morton, J. *The Natural History of Northamptonshire*, 1712.
Muhlfeld, H. B. (ed.) *A Survey of the Manor of Wye*, New York, 1933.
Müller-Wille, W. *Die Ackerfluren im Landesteil Birkenfeld und ihre Wandlungen seit dem 17. und 18. Jahrhundert*, Beiträge zur Landeskunde der Rheinlande, Veröffentlichungen des Geographischen Instituts der Universität Bonn, 2. Reihe, Heft 5, Bonn, 1936.

—— 'Das Rheinische Schiefergebirge und seine kulturgeographische Struktur und Stellung', *Deutsches Archiv für Landes- und Volksforschung*, vi, 1942.
—— 'Langstreifenflur und Drubbel: ein Beitrag zur Siedlungsgeographie Westgermaniens', ibid., viii, 1944.
Mundy, L. M. (ed.) *East Anglian Studies*, Cambridge, 1968.
Musset, L. 'Observations sur l'ancien assolement biennal du Roumois et du Lieuvin', *Annales de Normandie*, ii, 1952.

Naish, M. C. 'The Agricultural Landscape of the Hampshire Chalklands, 1700–1840', typescript thesis, M.A., London University, 1960.
Nef, J. U. *The Rise of the British Coal Industry*, 2 vols, 1932.
Neilson, N. (ed.) *A Terrier of Fleet, Lincolnshire, from a manuscript in the British Museum*, British Academy, Records of Social and Economic History vol. iv, 1920.
—— (ed.) *Cartulary and Terrier of Bilsington, Kent*, idem, vol. viii, 1928.
—— *Economic Conditions on the Manors of Ramsey Abbey*, Philadelphia, 1899.
Nettels, C. P. *The Roots of American Civilization*, New York, 1938.
Newton, K. C. *Thaxted in the Fourteenth Century*, Chelmsford, 1960.
Nielsen, A. *Dänische Wirtschaftsgeschichte*, Jena, 1933.
Norden, J. *Speculi Britanniae Pars: An Historical and Chorographical Description of the County of Essex* (1594) ed. H. Ellis, Camden Society, [ix], 1840.
—— *The Surveiors Dialogue*, 1618.
Notestein, W., Relf, F. H. and Simpson, H. (eds) *Commons Debates 1621*, 7 vols, New Haven, 1935.
Nourse, T. *Campania Foelix*, 1700.

Orwin, C. S. and C. S. *The Open Fields*, Oxford, 1938.
Osgood, H. L. *The American Colonies in the Seventeenth Century*, Gloucester, Mass., 1957.
Otway-Ruthven, J. 'The Organisation of Anglo-Irish Agriculture in the Middle Ages', *Journal of the Royal Society of Antiquaries of Ireland*, lxxxi, 1951.
Owen, G. *The Description of Penbrokshire*, ed. H. Owen, 4 pts, Cymmrodorion Record Series, no. 1, 1892–1936.

Page, F. M. *The Estates of Crowland Abbey*, Cambridge, 1934.
—— (ed.) *Wellingborough Manorial Accounts*, Northamptonshire Record Society, viii, 1936.
Palin, W. 'Farming of Cheshire', *Journal of the Royal Agricultural Society of England*, v, 1844.
Palmer, A. N. *The Town, Fields and Folk of Wrexham in the time of James I*, Wrexham and Manchester, n.d.
Palmer, A. N. and Owen, E. *A History of the Ancient Tenures of Land in North Wales and the Marches*, p.p. 1910.
Palmer, W. M. *A History of the Parish of Borough Green, Cambridgeshire*, Cambridge Antiquarian Society, Octavo Publications, liv, 1939.
Pape, T. *Newcastle-under-Lyme in Tudor and Early Stuart Times*, Manchester, 1938.
Parain, C. 'Travaux récents sur l'histoire rurale de Danemark', *Annales de Normandie*, ii, 1952.

Passmore, J. B. *The English Plough*, Oxford, 1930.
Payne, F. G. 'The Plough in Ancient Britain', *Archaeological Journal*, civ, 1948 (1947).
Pelham, R. A. 'The Agricultural Geography of the Chichester Estates in 1388', *Sussex Archaeological Collections*, lxxviii, 1937.
Petford, A. J. 'The Progress of Enclosure in Saddleworth 1025–1834', *Transactions of the Lancashire and Cheshire Antiquarian Society*, lxxxiv, 1987.
Phillimore, W. P. W. and Fry, G. S. (eds) *Abstracts of Gloucestershire Inquisitiones Post Mortem in the Reign of King Charles the First*, 3 pts, Index Library, British Record Society, 1893.
Phillpotts, B. S. *Kindred and Clan in the Middle Ages and After*, Cambridge, 1913 and New York, 1974.
Pierce, T. J. 'Agrarian Aspects of the Tribal System in Medieval Wales', in *Géographie et Histoire Agraires*, Annales de l'Est, q.v.
Pirenne, H. *Economic and Social History of Medieval Europe*, 1936.
Pitt, W. *A General View of the Agriculture of the County of Leicester*, 1809.
—— *A General View of the Agriculture of the County of Northampton*, 1809.
—— *A General View of the Agriculture of the County of Stafford*, 1794 and 1796.
Planhol, X. de, 'Essai sur la genèse du paysage rural de champs ouverts', *Géographie et Histoire Agraires*, Annales de l'Est, q.v.
Pliny the Elder, *Naturalis Historiae*.
Plot, R. *The Natural History of Oxfordshire*, Oxford, 1677.
—— *The Natural History of Staffordshire*, Oxford, 1686.
Pomeroy, W. *A General View of the Agriculture of the County of Worcester*, 1794.
Poole, A. L. *From Domesday to Magna Carta*, Oxford, 1955.
—— *The Obligations of Society in the Twelfth and Thirteenth Centuries*, Oxford, 1946.
Porter, J. 'Waste Land Reclamation in the Sixteenth and Seventeenth Centuries', *Transactions of the Historic Society of Lancashire and Cheshire*, cxxvii, 1978 (1977).
Postgate, M. R. 'The Field Systems of Breckland', *Agricultural History Review*, x, 1962.
—— 'Field Systems of East Anglia', in Baker and Butlin, q.v.
—— 'Historical Geography of Breckland, 1600–1850', typescript thesis, M.A., London University, 1960.
Potter, S. P. *A History of Wymeswold*, 1915.
Powicke, M. *The Thirteenth Century: 1216–1307*, Oxford, 1962.
Prestwich, M. *The Three Edwards: war and state in England 1272–1377*, 1980.
Priest, St J. *A General View of the Agriculture of Buckinghamshire*, 1813.
Purvis, J. S. (ed.) *Select XVI Century Causes in Tithe from the York Diocesan Registry*, Yorkshire Archaeological Society Record Series, cxiv, 1949 (1947).

Raftis, J. 'The East Midlands', in Finberg and Thirsk, q.v., ii.
Raine, J. (ed.) *The Priory of Hexham*, 2 vols, Surtees Society, xliv, xlvi, 1864–5.
Raistrick, A. *Malham and Malham Moor*, Clapham via Lancaster, 1947.
Ratcliffe, S. C. and Johnson, H. C. (eds) *Warwick County Records, vol. iv: Quarter Sessions Order Book*, Warwick, 1938.
Ravensdale, J. R. 'Landbeach in 1549: Ket's Rebellion in Miniature', in L. M. Mundy, q.v.

—— *Liable to Floods: a village landscape on the edge of the fens A.D. 450–1850*, Cambridge, 1974.
Richardson, H. (ed.) *Court Rolls of the Manor of Acomb*, 2 vols, Yorkshire Archaeological Society Record Series, cxxxi, cxxxvi, 1969–78 (1968–75).
Richeson, A. W. *English Land Measuring to 1800: instruments and practices*, Society for the History of Technology and the Massachusetts Institute of Technology, Cambridge, Mass. and London, 1966.
Rider-Haggard, H. *Rural England*, 2 vols, 1906.
Ritchie, J. (ed.) *Reports of Cases decided by Francis Bacon in the Court of Chancery (1617–21)*, 1932.
Roberts, B. K. 'Field Systems of the West Midlands', in Baker and Butlin, q.v.
Robertson, A. J. (ed.) *Anglo-Saxon Charters*, Cambridge, 1956.
Robinson, D. J., Salt, J. and Phillips, A. D. M. 'Strip Lynchets in the Peak District', *North Staffordshire Journal of Field Studies*, ix, 1969.
Robinson, G. T. *Rural Russia under the Old Régime*, 1932.
Roden, D. 'Field Systems of the Chiltern Hills and their Environs', in Baker and Butlin, q.v.
—— 'The Fragmentation of Farms and Fields in the Chiltern Hills: 13th century and later', *Medieval Studies*, xxxi, 1969.
Roderick, A. J. 'Open-Field Agriculture in Herefordshire in the late Middle Ages', *Transactions of the Woolhope Naturalists' Field Club, Herefordshire*, iii, 1949–51.
Root, H. L. *Peasants and King in Burgundy: agrarian foundations of French absolutism*, Berkeley, Calif., Los Angeles and London, 1987.
Ross, C. D. and Devine, M. (eds.) *The Cartulary of Cirencester Abbey*, 3 vols, London and Oxford, 1964–77.
Rowe, J. *Cornwall in the Age of the Industrial Revolution*, Liverpool, 1953.
Rowley, T. (ed.) *The Origins of Open-Field Agriculture*, 1981.
Rowse, A. L. *Tudor Cornwall*, 1941.
Ruddle, C. S. 'Notes on Durrington', *Wiltshire Archaeological Magazine*, xxxiii, 1903–4.
Ruston, A. G. and Witney, D. *Hooton Pagnell*, 1934.
Ryder, M. L. and Stephenson, S. K. *Wool Growth*, London and New York, 1968.
Rylands, J. P. (ed.) *Lancashire Inquisitions returned into the Chancery of the Duchy of Lancaster, Stuart Period*, 3 pts, The Record Society for Lancashire and Cheshire, ii, xvi, xvii, 1880–8.

Saltmarsh, Col. 'Some Howdenshire Villages', *Transactions of the East Riding Archaeological Society*, xiii, 1907 (1906–7).
Saltmarsh, J. and Darby, H. C. 'The Infield–Outfield System on a Norfolk Manor', *Economic History*, iii, 1934–7.
Savine, A. *English Monasteries on the Eve of the Dissolution*, Oxford Studies in Social and Legal History, i, Oxford, 1909.
Schröder, K. H. 'Die Flurformen Württemberg-Hohenzollern und ihre neuzeitliche Umgestaltung', *Raumforschung und Raumordnung*, v, 1941.
Schröder-Lembke, G. 'Die Entstehung und Verbreitung der Mehrfelderwirtschaft in Nordostdeutschland', *Zeitschrift für Agrargeschichte und Agrarsoziologie*, ii, 1954.

—— 'Römische Dreifelderwirtschaft?', ibid., xi, 1963.
—— 'Wesen und Verbreitung der Zweifelderwirtschaft im Rheingebiet', ibid., vii, 1959.
Scofield, E. 'The Origin of Settlement Patterns in Rural New England', *Geographical Review*, [xxviii], 1938.
Scrope, G. P. *History of the Manor and Ancient Barony of Castle Combe in the County of Wiltshire*, sine loco, 1852.
Scrutton, T. E. *Commons and Common Fields*, Cambridge, 1887.
Sée, H. *Esquisse d'une histoire du régime agraire en Europe aux XVIIIe et XIXe siècles*, Paris, 1921.
Seebohm, F. *The English Village Community*, 1883.
Sheppard, J. A. 'Field Systems of Yorks.' in Baker and Butlin, q.v.
—— *The Origins and Evolution of Field and Settlement Patterns in a Herefordshire Manor, Marden*, Queen Mary College, London, Department of Geography Occasional Paper 15, 1979.
Shorter, A. H. 'Ancient Fields in Manaton Parish, Dartmoor', *Antiquity*, xii, 1938.
—— 'Field Patterns in Brixham Parish', *Report and Transactions of the Devonshire Association*, civ, 1972.
Simkhowitsch, W. G. *Die Feldgemeinschaft in Russland: ein Beitrag zur Sozialgeschichte und zur Kenntnis der gegenwärtigen wirtschaftlichen Lage des russischen Bauernstandes*, Jena, 1898.
Simpson, A. 'The East Anglian Foldcourse: some queries', *Agricultural History Review*, vi, 1958.
—— *The Wealth of the Gentry 1540–1660*, Cambridge, 1961.
Sinclair, J. *Agricultural Hints*, n.d.
Singleton, J. 'The Influence of Geographical Factors on the Development of the Common Fields of Lancashire', *Transactions of the Historic Society of Lancashire and Cheshire*, cxv, 1964 (1963).
Sion, J. *Les Paysans de la Normandie Orientale*, Paris, 1909.
Skipp, V. 'The Evolution of Settlement and Open-Field Topography in North Arden down to 1300', in T. Rowley, q.v.
Slack, W. J. (ed.) *The Lordship of Oswestry*, Shrewsbury, 1951.
—— 'The Open Field System of Agriculture', *Transactions of the Carodoc and Severn Valley Field Club*, x, 1937.
Slater, G. (ed.) *Economic Studies, vol. i, Some South Indian Villages*, 1918.
—— *The English Peasantry and the Enclosure of Common Fields*, 1907.
Slee, A. H. 'The Open Fields of Braunton', *Report and Transactions of the Devonshire Association*, lxxxiv, 1952.
Slicher van Bath, B. H. 'Economic and Social Conditions in the Frisian Districts from 900 to 1500', *A. A. G. Bijdragen*, xiii, 1965.
—— *Mensch en Land in de Middeleeuwen*, 2 vols, Academisch Proefschrift, Assen, [1943].
—— 'Studien betreffende de agrarische Geschiedenis van de Veluwe in de Middeleeuwen', *A. A. G. Bijdragen*, xiii, 1965.
—— *The Agrarian History of Western Europe A.D. 500–1850*, 1963.
—— 'The Rise of Intensive Husbandry in the Low Countries', in J. S. Bromley and E. H. Kossmann (eds) *Britain and the Netherlands*, 1960.

Smith, J. *Chronicon Rusticum-Commerciale or Memoirs of Wool etc.*, 2 vols, 1747.
Smith (Smyth), J. of Nibley, see Smyth.
Smith, R. E. F. *The Enserfment of the Russian Peasantry*, Cambridge, 1968.
Smith, T. C. *The Agrarian Origins of Modern Japan*, Stanford, Calif., 1959.
Smyth (Smith), J. *A Description of the Hundred of Berkeley*, Berkeley Manuscripts iii, ed. J. Maclean, Gloucester, 1885.
—— *The Lives of the Berkeleys*, 2 vols, Berkeley Manuscripts i and ii, ed. J. Maclean, 1883.
Sombart, W. *Die Deutsche Volkswirtschaft im Neunzehnten Jahrhundert und im Anfang des 20. Jahrhunderts*, Berlin, 1927.
Spratt, J. 'Agrarian Conditions in Norfolk and Suffolk, 1600–1650', typescript thesis, M.A., London University, 1935.
Spufford, M. *A Cambridgeshire Community*, Leicester, 1965.
—— *Contrasting Communities: English villagers in the sixteenth and seventeenth centuries*, Cambridge, 1974.
—— 'Rural Cambridgeshire 1520–1680', typescript thesis, M.A., Leicester University, 1962.
Stark, A. *The History and Antiquities of Gainsborough*, 1843.
Steensberg, A. 'Sula: an ancient term for the wheel plough in northern Europe', *Tools and Tillage*, iii, 1976–9.
Steinbach, F. 'Geschichtliche Siedlungsformen in der Rheinprovinz: Angaben der Siedlungsgeschichte in der Rheinprovinz', *Zeitschrift des Rheinischen Vereins*, xxx, 1937.
Steinbach, F. and Becker, E. *Geschichtliche Grundlagen der kommunalen Selbstverwaltung in Deutschland*, Bonn, 1932.
Stenton, F. M. *Anglo-Saxon England*, Oxford, 1971.
—— (ed.) *Documents Illustrative of the Social and Economic History of the Danelaw*, British Academy, 1920.
Stocks, H. (ed.) *Records of the Borough of Leicester 1603–88*, Cambridge, 1923.
Stocks, J. E. (ed.) *Market Harborough Parish Records 1531–1837*, 1926.
Stone, T. *A Review of the Corrected Agricultural Survey of Lincolnshire by Arthur Young Esquire*, 1800.
Straker, E. *Wealden Iron*, 1931.
Straton, C. R. (ed.) *Survey of the Lands of William, first Earl of Pembroke*, 2 vols, Roxburghe Club, 1909.
Sylvester, D. 'A Note on Medieval Three-course Arable Systems in Cheshire', *Transactions of the Historic Society of Lancashire and Cheshire*, cx, 1959 (1958).
—— 'Rural Settlement in Cheshire', ibid., ci, 1950 (1949).
—— 'The Open Fields of Cheshire', ibid., cviii, 1957 (1956).
—— 'Settlement Patterns in Rural Flintshire', *Flintshire Historical Society's Publications*, xv, 1954–5.
—— 'The Common Fields of the Coastland of Gwent', *Agricultural History Review*, vi, 1958.
—— *The Rural Landscape of the Welsh Borderlands: a study in historical geography*, 1969.

Tacitus, C. C. *Germania.*
Tate, W. *The Parish Chest*, Cambridge, 1951.

Tawney, R. H. *The Agrarian Problem in the Sixteenth Century*, 1912.
Textor, L. E. *Land Reform in Czechoslovakia*, 1923.
Thallóczy, L. von (ed.) *Illyrisch–Albanische Forschungen*, 2 vols, Munich and Leipzig, 1916.
Thirsk, J. 'The Common Fields' in R. H. Hilton, *Peasants, Knights and Heretics*, q.v.
Thomas-Stanford, C. (ed.) *An Abstract of the Court Rolls of the Manor of Preston (Preston Episcopi)*, Sussex Record Society, xxvii, 1921.
Thompson, J. W. *Feudal Germany*, Chicago, 1928.
Thornton, G. A. *A History of Clare, Suffolk*, Cambridge, 1930.
—— 'A Study in the History of Clare, Suffolk', *Transactions of the Royal Historical Society*, 4th series, xi, 1928.
Titow, J. Z. *English Rural Society 1200–1350*, 1969.
—— 'Medieval England and the Open Field System' in R. H. Hilton, *Peasants, Knights and Heretics*, q.v.
Trewartha, G. T. 'Types of Rural Settlement in Colonial America', *Geographical Review*, xxxvi, 1946.
Tupling, G. H. *The Economic History of Rossendale*, Chetham Society, new series, lxxxvi, 1927.
Turner, G. *A General View of the Agriculture of the County of Gloucester*, 1794.
Tusser, T. *Five Hundred Points of Good Husbandry*, ed. G. Grigson, Oxford, 1984.
Twemlow, F. R. 'The Manor of Tyrley in the County of Stafford', *Collections for the History of Staffordshire*, 3rd series, 1948 (1945–6).

Vancouver, C. *A General View of the Agriculture of the County of Cambridge*, 1794.
—— *A General View of the Agriculture of the County of Devon*, 1808.
—— *A General View of the Agriculture of the County of Essex*, 1795.
Vanzetti, C. 'L'Utilisation du sol', in Guarducci, q.v.
Venard, M. *Bourgeois et paysans au XVIIe siècle*, Ecole Pratique des Hautes Etudes, VIe section, Centre de Recherches Historiques: Les Hommes et la Terre, no. III, Paris, 1957.
Venn, J. A. *Foundations of Agricultural Economics*, Cambridge, 1933.
—— 'The Economy of a Norfolk Parish in 1783 and at the Present Time', *Economic History*, i, 1929.
Verhulst, A. E. 'Probleme der mittelalterlichen Agrarlandschaft in Flandern', *Zeitschrift für Agrargeschichte und Agrarsoziologie*, ix, 1961.
Vernadsky, G. *Kievan Russia*, vol. ii of History of Russia, ed. G. Vernadsky and M. Karpovich, New Haven, 1948.
Verney, F. P. *Memoirs of the Verney Family during the Civil War*, 3 vols, 1892–4.
Vinogradoff, P. *English Society in the Eleventh Century*, Oxford, 1908.
—— *The Growth of the Manor*, 1904.
—— *Villainage in England: essays in English medieval history*, Oxford, 1892.
Vinogradoff, P. and Morgan, F. (eds) *Survey of the Honour of Denbigh 1334*, British Academy, Records of Social and Economic History, vol. i, 1914.
Vivian, S. P. *The Manor of Etchingham cum Salehurst*, Sussex Record Society, liii, 1953.

Wake, J. 'Communitas Villae', *English Historical Review*, xxxvii, 1922.

Walker, D. *A General View of the Agriculture of the County of Hertford*, 1795.
Wallich, H.C. *Mainsprings of the German Revival*, New Haven, 1955.
Wanklyn, H.G. *The Eastern Marchlands of Europe*, 1941.
Warriner, D. (ed.) *Contrasts in Emerging Societies*, 1965.
—— *Economics of Peasant Farming*, 1939.
—— *Land Reform and Development in the Middle East*, 1962.
—— 'Some Controversial Issues in the History of Agrarian Europe', *Slavonic and East European Review*, xxxii, 1953.
Wartmann, H. (ed.) *Urkundenbuch der Abtei Sanct Gallen*, 2 vols, Zurich, 1863–6.
Waylen, J. *A History Military and Municipal of the Town (otherwise called the city) of Marlborough*, 1844.
Webb, S. and B. *English Local Government: the Manor and the Borough*, pt i, 1908.
Wedge, J. *A General View of the Agriculture of the County of Warwick*, 1794.
Weulersse, J. *Paysans de Syrie et du Proche-Orient*, Paris, 1946.
White, K.D. *Agricultural Implements of the Roman World*, Cambridge, 1967.
White, L. *Medieval Technology and Social Change*, Oxford, 1962.
Whitelock, D. (ed.) *Anglo-Saxon Wills*, Cambridge, 1930.
—— *The Beginnings of English Society*, Harmondsworth, 1952.
Whiteman, R.J. (ed.) *Hexton: a parish survey*, p.p. 1936.
Whittington, G. 'Field Systems of Scotland', in Baker and Butlin, q.v.
Wightman, W.E. 'Open Field Agriculture in the Peak District', *Derbyshire Archaeological Journal*, lxxxi, 1961.
Wilkerson, J.C. (ed.) *John Norden's Survey of Barley, Hertfordshire, 1593–1603*, Cambridge Antiquarian Records Society, ii, 1973.
Willan, T.S. (ed.) *A Tudor Book of Rates*, Manchester, 1962.
Willis, D. (ed.) *The Estate Book of Henry de Bray of Harlestone, Co. Northants. (c. 1289–1340)*, Royal Historical Society, Camden Series, xxvii, 1916.
Willson, D.H. (ed.) *The Parliamentary Diary of Robert Bowyer 1606–7*, Minneapolis, 1931.
Wimbledon Common Committee, *Extracts from the Court Rolls of the Manor of Wimbledon*, 1866.
Woodward, D. (ed.) *The Farming and Memoranda Books of Henry Best of Elmswell 1642*, British Academy, [1984].
Woodward, D.M. *The Trade of Elizabethan Chester*, Hull, 1970.
Worlidge, J. *Systema Agriculturae*, 1669.
Wührer, K. *Beiträge zur ältesten Agrargeschichte des germanischen Nordens*, Jena, 1935.

Yelling, J.A. *Common Field and Enclosure in England 1450–1850*, 1977.

Zaleska, M.K. 'L'Ancien Morcellement des champs avant la séparation au XIXe siècle dans la Poméranie de Gdansk', in *Géographie et Histoire Agraires*, Annales de l'Est. q.v.
Zimányi, V. 'La Formation des monocultures et ses conséquences ecologiques en Hongrie', in Guarducci, q.v.
Żytkowicz, L. 'Les Transformations du paysage agraire polonais avant le XIXe siècle', in Guarducci, q.v.
—— 'The Peasant's Farm and the Landlord's Farm in Poland from the Sixteenth to the Mid-Eighteenth Century', *Journal of European Economic History*, i, 1972.

Summary of select manuscript sources

Public Record Office

Court of Chancery: Proceedings, series i, James I
Court of Common Pleas: Recovery Rolls
Court of Exchequer:
  Augmentation Office: Miscellaneous Books; Parliamentary Surveys
  King's Remembrancer: Depositions by Commission; Miscellaneous Books; Special Commissions
  Land Revenue: Miscellaneous Books
  Treasury of Receipt: Books
Court of Requests: Proceedings
Court of Star Chamber: Proceedings
Court of Wards and Liveries: Feodaries' Surveys
Duchy of Lancaster: Court Rolls; Miscellaneous Books; Rentals and Surveys; Special Commissions
Privy Council: Registers
Special Collections: Court Rolls, General Series; Maps and Plans; Rentals and Surveys; State Papers, Domestic

British Library

Additional Charters
Additional Manuscripts
Additional Rolls
Cottonian Manuscripts: Augustus, Claudius, Julius, Nero, Tiberius, Vespasian
Egerton Charters
Egerton Manuscripts
Harleian Manuscripts
Sloane Manuscripts
Stowe Charters
Stowe Manuscripts

Bedfordshire Record Office

AD.1060; BS.1276; C.364; H.A.5/1; L.4/333; L.26/10; L.26/280–1; L.26/563; T.W.10/2/8,9; X.69/6

Berkshire Record Office

D/EBt/E.28; D/EC/M.26, M.151; D/EHy/M.32; D/EPb/M.3

Birmingham Reference Library

General Collection: 168002; 168162–3; 32480; 344741–2; 378173; 382959; 437912; 478550; 499837; 505455

Bodleian Library

Department of Western Manuscripts: Aubrey MSS, vols 2, 3; Hearne's Diaries 158–9

Boughton House

Montagu Manuscripts: Compota Omnium et Singulorum Officiariorum et Ministorum (Sir Edward Montagu's Account Book)

Bowood House

Survey of Bremhill 1629

Bristol Archives Office

MS 4490

Buckinghamshire Record Office

Acquisition 35/39 Bye-laws and Regulations for Padbury Fields 1779; Swanbourne Field Agreement 1748; D/MH 28/1; Ex-Museum MSS, Court Rolls and Files: Emberton 616/43; Ivinghoe IM.6/1, IM.6/12 (Court Book); Long Crendon 10/48; Pitstone P.24/1, 2, 3; Stone 452/39; Taplow 155/21; West Wycombe W.10/1

Cambridgeshire Record Office

L.1/8,9,10 Court Rolls, Bassingbourne Richmond; L.1/20 Court of Survey, Fen Ditton 30 Sept. 1672; L.1/112 Court Rolls, Long Stanton; L.1/182 Court Rolls, Willingham; L.19/17–21 Court Rolls, Haslingfield; L.64 Court Rolls, Chatteris Foxton, Wimbish Foxton and Foxton Bury; L.88/4 Court Files, Caxton; R.51.29.1 Court Rolls, Ickleton manors

Cheshire Record Office

Court of the Vicar General of the Chancellor of the Diocese of Chester in the Episcopal Consistory of Chester and in the Rural Deaneries of the Archdeaconry: Original Wills and Administrations; Nedeham and Kilmorey Collection; Vernon (Warren) Collection

Chetham Library, Manchester

Adlington Manuscripts

Coventry Record Office

City Records

Deene House (Deene Park)

Brudenell Manuscripts

Devizes Museum

William Gaby His Booke, 1656

Dorset Record Office

D.10 (Weld Collection) E.103; D.39/H.2 Puddletown Court Book; D.54 Surveys, Rentals and Terriers of Manors of Lady Margaret Arundel

Essex Record Office

D.DL M.81; D.DP E.25–6, M.186, M.890; D.Vm 20

Gloucester Reference Library

Local Collection: 16062

Gloucestershire Record Office

Consistory Court of Gloucester: inventories; D.127/608; D.158 M.1, M.7, M.24; D.184/M.7, M.9, M.18–20, M.24; D.326/E.1; D.444/M.1, M.2; D.445/M.12; D.745 Bisley Court Files

Hertfordshire Record Office

Archdeaconry of St Albans: inventories; Ashridge Collection; Broxborne Bury Collection; Cashiobury and Gape Collection; Dimsdale Collection; Gorhambury Collection; Moulton Collection; Pym Collection; Miscellaneous Manuscripts

Lancashire Record Office

DD.F, 158, 168, 1649; DD.K, 1451/1, 1452/1, 1456/2, 1470/1, 1506/2, 1526/5, 1542/2; DD.X.80/1

Leicester Museum

MS 35/29/372

Leicester Town Hall

Borough Records: BR.II.8.75

Leicestershire Record Office

Beaumanor Collection (DE.10); Commissary of the Bishop of Lincoln in the Archdeaconry of Leicester and the Court of the Archdeacon: Original Wills and Administrations, Inventories; Museum Manuscripts

Lichfield Joint Record Office

Peculiar Courts of Alrewas and Weeford, Colwich, Gnossal, Hansacre and Armitage, and Penkridge: Original Wills

Lincolnshire Archives Office

Monson Deposit: Newton Papers; Heneage of Hainton Collection

Liverpool Record Office

Norris Papers

Manchester Library

Miscellaneous Deeds

Middlesex Record Office

Accession no. 249/875

Norfolk and Norwich Record Office

Dean and Chapter Manuscripts; Episcopal Consistory Court of Norwich: inventories; Norwich Library Manuscripts

Northamptonshire Record Office

Aynho Court Rolls; Consistory Court of Peterborough: Bonds and Inventories; Administrations and Inventories; Daventry Collection; Finch-Hatton Collection; Fitzwilliam (Milton) Collection; Grafton Collection; Isham (Lamport) Collection; Miscellaneous Ledgers; Montagu Collection; Old Parish Records; Rushden Court Rolls; Sotheby (Ecton) Collection; Westmorland Collection; YZ Collection

Oxfordshire Record Office

DIL.II.a.2; DIL.II.w.75; DIL.IV.a.89; Wi.x.34; Dashwood Collection: VIII/xxxiii, xxxiv; Miscellaneous M.I/1

Shakespeare's Birthplace Library, Stratford-upon-Avon

Coughton (Throckmorton) Collection; Gregory-Hood Collection; Manorial Documents and Court Rolls; Stoneleigh (Leigh) Collection; Stratford-on-Avon Collection; Trevelyan Collection; Willoughby de Broke Collection

Shropshire Record Office

MS 167/43

Somerset Record Office

DD/CN (Accession C.168 box 2 no. 26) Court Book of John Francis Esq. 1573–87; DD/HP Transcript of articles of agreement between lords farmers and tenants of West Buckland 1634; DD/MI (Mildmay Coll.) Accession no. C.186; DD/SAS Aller custumal 1653; DD/SP Accession no. H.62; DD/X/GB Combe Survey Book 1704

Staffordshire Record Office

Hatherton Manuscripts (D.260); Madeley Holme Court Rolls (D.1750); Hand Morgan Collection: Chetwynd Manuscripts, file N: Court Rolls of Churcheaton; file O: Court Rolls of Ingestre cum membris; Shenstone Court Rolls

Suffolk Record Office (Ipswich)

51.2.12; 51.10.17.3; V5.18.10.1; V5.23.2.1; V11.2.1.1

Surrey Record Office

Survey of Great Bookham 1614 (34/3)

Urchfont Manor House

Copies of Urchfont Enclosure Act and Award, 1789, 1793

Warwickshire Record Office

Manorial Records, M.R.21 Court Rolls of Berkswell

Wilton House

Papers relating to Various Parishes: Court Book of Various Parishes 1633–4; Court Books of Stanton and of Stoke Farthing beginning 1724; Court Books of Ebbesbourne and of Foffont beginning 1742–3; Court Rolls 1666–89; Court Rolls of Manors 1689–1754 (1690–1723); Surveys of Manors in Co. Wilts.; Surveys of Manors 1631; Surveys of Manors 1632–3; Survey of Flambston 1631; Survey of the Manor of Alvediston 1706 and 1758

Wiltshire Record Office

Accession 7, Enford Court book; Accession 212B: Broad Hinton B.H.8; Grittleton Court Rolls; Accession 283; Liber Supervisus maneriorum de Amisburie Erles and Amesbury Pryorye (Survey of Amesbury with Baicliffe); Amesbury Survey Book 1574; Survey of Amesbury 1635; Amesbury Miscellaneous Papers and Documents: Billet land cum aliis; Enclosure Awards: Fovant, Ebbesbourne Wake, Broadchalke, Bowerchalke, Alvediston, Bishopstone, and Fyfield 1792; Ex Salisbury Diocesan Registry: Miscellaneous Wills, Administrations and Inventories; Hyde Family Documents; Keevil and Bulkington Manuscripts; Savernake Collection

Worcestershire Record Office (Hereford and Worcester)

B.A.54 Kempsey Court Rolls; B.A.68 Woodmanton Valuation 1611, Woodmanton lease covenant memorandum 1612; B.A.351 bundle 8 no. 1, Pensham terrier 1794; B.A.494 Particulars of Morton Underhill 1648 and Great Comberton circa 1660; Episcopal Consistory Court of Worcester: Original Wills, Administrations and Inventories.

Index of persons

Index of places

Glossarial index of subjects
(glossarial entries italicised)